FREE WILL
and
HUMAN LIFE

ALAN E. JOHNSON

Philosophia Publications
Pittsburgh, Pennsylvania

Copyright © 2021 by Alan E. Johnson
All rights reserved. Published 2021

For permissions or other information,
contact Alan E. Johnson at
https://alanjohnson.academia.edu/contact.

Print ISBN: 978-0-9701055-3-0
Digital ISBN: 978-0-9701055-4-7
Library of Congress Control Number: 2021914728

*Cover image courtesy of James Wheeler,
"Two Paths," https://www.souvenirpixels.com
(adaptation used with permission)*

Published by
Philosophia Publications
301 South Hills Village Drive
Suite LL200-112
Pittsburgh, Pennsylvania 15241
USA

http://www.PhilosophiaPublications.com

To the Memory of My Parents

CONTENTS

Preface v
Introduction 1
Chapter 1   Arguments against Free Will 7
Chapter 2   Arguments for Free Will 48
Chapter 3   Free Will and Human Nature 88
Conclusion 102
Notes 104
Bibliography 134
Index 149
About the Author 153

PREFACE

This is the first book of a planned three-volume philosophical trilogy on free will, ethics, and political philosophy. It began as a projected first chapter of a book on ethics that I was preparing in the 2010s to replace my 2000 book *First Philosophy and Human Ethics: A Rational Inquiry*. The 2000 book assumed, without investigating the question, that humans have free will. In reexamining this premise, I became aware of a vast academic literature arguing that free will is a mere illusion. I take it as axiomatic that the question whether a philosophical approach to human ethics is even possible depends on a preliminary question whether humans have free will to formulate and apply ethical principles.

As I researched and wrote the chapter on free will for the replacement book on ethics, I came to realize that my discussion could not be confined to the limitations of a chapter. Accordingly, I put the ethics book on hold in order to prepare the present book on free will. I was then substantially diverted from this task by preparing a first and then second edition of my book on the U.S. Electoral College, following the presidential elections of 2016 and 2020, respectively. I have finally, however, completed *Free Will and Human Life*. The ethics book, to which I will now return, references the present writing for the benefit of those who wish to consider the question of free will before proceeding to a study of ethics.

The Introduction, below, provides an overview of the subject matter of this book together with my definition of free will. I then discuss arguments against free will (Chapter 1), arguments for free will (Chapter 2), and my own conclusions about free will (Chapter 3). The thinkers discussed in Chapters 1 and 2 are representative rather than exclusive. Many others have written on these questions, but time and space do not permit an exhaustive treatment of all such opinions.

I wish to thank my wife, Mimi Lindauer, for reviewing and commenting on drafts of this book for typos, grammar, and style and for our numerous substantive discussions regarding the question of free will over the years.

Thanks also to astrophysicist-philosopher Robert O. (Bob) Doyle and independent philosopher Robert Hanna for reviewing and commenting on some drafts of portions of this work and also for their publications and their email communications with me regarding free will. I appreciate their substantive comments, which caused me to think further about some relevant issues, though my own analysis differs from each of theirs in certain details. Robert Hanna also made some stylistic suggestions, many of which I have adopted. Of course, I alone am responsible for the final product.

This book is limited to the question of free will. When I discuss or cite other writers, my agreement or disagreement with them should not be construed to extend to issues other than those being immediately addressed. My forthcoming works on ethics and

PREFACE

political philosophy will separately address matters relevant to those topics.

When I am discussing a particular work in the present book, I sometimes put cited page numbers of that writing in textual parentheses instead of in endnotes. In all such cases, these citations refer to the book or paper last mentioned in the text.

The page numbers of the paperback edition of this book are inserted in italicized braces (*{}*) at the appropriate locations of the ebook editions. The page numbers identified in the Index of the ebooks refer to these italicized page numbers, and that Index is accordingly identical to the Index in the paperback.

I will post errata and supplemental comments, if any, at
https://chicago.academia.edu/AlanJohnson/Books,-Book-Excerpts,-and-Errata-Supp-Comments.

Alan E. Johnson
July 25, 2021

INTRODUCTION

If everything, whether organic or inorganic, operates according to a predetermined sequence of cause and effect—if we are mere robots acting out a script written by God or Nature at or before the beginning of time—then what will be will be, and free will does not exist. Humans are playthings of the gods—or, to use a more updated reference, living exemplars of a preprogrammed artificial intelligence. In the words of Macbeth, "Life's but a walking shadow, a poor player, that struts and frets his hour upon the stage, and then is heard no more. It is a tale told by an idiot, full of sound and fury signifying nothing."[1]

We have just described the predeterministic theory of classical, Newtonian physics.[2] This type of rejection of free will is similar in important respects to theological predeterminism.[3] But the discovery in the twentieth century of quantum physics suggested—over the objection of Albert Einstein, who famously said that God does not play dice with the universe—that the predeterministic theory is incorrect. Quantum mechanics shows that physics is grounded on indeterministic phenomena. As a result of the law of large numbers, indeterminism averages out to produce "laws" governing matter and energy. Quantum physicists argue that physical phenomena are ultimately governed by probability, not determinism. In light of the extensive empirical verification of quantum mechanics,

INTRODUCTION

most, though not all, physicists and other scholars now accept these principles.[4]

Such facts have given rise to another kind of attack on free will. This argument claims that, to the extent predeterminism does not rule, everything is random. If predeterminism does not govern us, then pure chance does.[5] These enemies of free will do not pause to consider the possibility that they are committing the fallacy of false dichotomy.[6] There may be a third option, which will be discussed later in this book.

Most people reject both the predeterminist and indeterminist objections out of hand. Our everyday experience seems to confirm that we have free will.[7] But opponents of free will condescendingly dismiss such "folk" wisdom (as they call it) as unsophisticated.[8] The word "unsophisticated" takes us back to the lifetime of Socrates (469–399 BCE) in ancient Greece. The root of this word is "sophist." The sophists were ancient purveyors of intellectual doctrines that challenged the commonsense views of ordinary people. Socrates famously cross-examined sophists and others claiming to have superior wisdom. Through his "Socratic method," he demonstrated that these intellectual luminaries did not, in fact, know what they claimed to know.[9]

This book examines fundamental questions concerning free will. Some scholars have identified the evolutionary development of free will from other life forms to humans.[10] The present discussion focuses specifically on **human** free will.

Astrophysicist-philosopher Bob (Robert O.) Doyle has made the following observation about contemporary debates among the academic proponents

INTRODUCTION

and opponents of free will: "Unfortunately, their works are full of a dense jargon defining (sometimes obscuring) subtle differences in their views—agent causation, event causation, non-occurrent causation, reasons as causes, intentions, undefeated authorization of preferences as causes, noncausal accounts, dual control, plurality conditions, conditions, origination, actual sequences and alternative sequences, source and leeway compatibilism, revisionism, restrictivism, semicompatibilism, and narrow and broad incompatibilism."[11] Doyle himself has nevertheless made a herculean effort to disentangle and clarify the many accounts of the (pre)determinists, the compatibilists, and the advocates of free will.[12] Although the discussion in the present book will make some basic distinctions, it will leave abstruse metaphysical speculation to the philosophy professors. We focus on important contributions to the free will debate with a view toward articulating the principal arguments rather than elaborating the disputational minutiae. We start with the surprisingly difficult task of providing an adequate and useful definition of free will.

DEFINING FREE WILL

Various definitions of "free will" have been proffered over the centuries and millennia. The definition used in this book is as follows: **"Free will" is the independent ability to make conscious decisions that are neither predetermined nor random.** This definition is similar but not identical to some other definitions of "free will." For example, Doyle defines free will as "a two-stage creative process in which a human or higher animal freely generates alternative possibilities, some caused

INTRODUCTION

by prior events, some uncaused, following which the possibilities are evaluated and one is 'willed,' i.e., selected or chosen for adequately determined reasons, motives, or desires."[13] Neuroscientist William R. (W. R.) Klemm defines the term as follows: "Free will occurs when a person makes a conscious choice from multiple options, none of which are predetermined or compelled.[14]

The word **"conscious"** in our definition does not necessarily mean fully conscious at the exact time of the event. Neuroscience and common sense both establish that one can consciously decide to learn and practice a behavior (for example, driving a car, flipping a signal light in a car, playing the piano, riding a bicycle, typing, playing tennis, acting ethically) that becomes habitual through repetition and may not be fully conscious at a time after it becomes automatic.[15] Additionally, consciousness of an event may be quickly forgotten when the time span of the consciousness is exceedingly short.[16] "Conscious" does, however, exclude nonconscious internal bodily operations such as molecular chemistry or physical neuronal processes.[17] And although subconscious influences may sometimes affect the rational exercise of free will, one may have or develop free will to overcome certain subconscious influences through deliberation and the acquisition of knowledge.

"Independent," "ability," and "random" are also important and necessary components of the definition of free will. Persons who have been systematically brainwashed or hypnotized do not have an **"independent"** ability.[18] Those who suffer from the more serious forms of psychosis or brain injury do not have the requisite **"ability"** for free will. A decision is **"random"** if it is dictated solely by chance. However, a decision is **not random** if the decider consciously chooses between alternatives by submitting the decision to a chance mechanism such as a coin toss.

The word **"predetermined"** is also essential to this definition. **Predeterminism** "is the idea that a strict causal determinism is true, with a causal chain of events back to the origin of the universe, and one possible future."[19] **Causal determinism** "is, roughly speaking, the idea that every event is necessitated by antecedent events and conditions together with the laws of nature."[20] **Determinism** "is the idea that everything that happens, including all human actions, is completely determined by prior events. There is only one possible future, and it is completely predictable in principle, . . . assuming perfect knowledge of the positions, velocities, and forces for all the atoms in the void."[21] These definitions create some very fine, arguably imperceptible distinctions. The difference appears to be that "predeterminism" goes back to the beginning of the universe, whereas "determinism" and "causal determinism" may go back only to some point **after** the Big Bang (interrupted by an interval of indeterminism or by God's supposed miraculous suspension of the laws of nature). In practice, these terms are often used interchangeably. In the present book, however, the term "predeterminism" (also called "universal natural determinism") is used to refer to allegedly inevitable cause and effect from the beginning of the universe to the present, whereas "determinism" refers to any kind of causal mechanism, whether predetermined or not.

INTRODUCTION

The constraints at issue in free will inquiry are those operating in the individual human brain, as distinguished from impediments imposed by other human beings, one's economic situation, the physical environment, or other circumstances. In the absence of predeterminism, any factual inability to realize one's thoughts in action does not negate free will. One can still decide, within one's own mind, the correct way of thinking and acting regardless of one's practical ability to execute such mental conclusions.[22]

OVERVIEW OF BOOK

Having defined free will, we now turn to arguments against (Chapter 1) and for (Chapter 2) free will, followed by my own perspective (Chapter 3). The views discussed in Chapters 1 and 2 are not exhaustive. Space (not to mention time) does not permit a treatment of all arguments that have been made against and for free will. I have selected those authors that I regard as most representative of the various positions on this subject.

CHAPTER 1

ARGUMENTS AGAINST FREE WILL

> I have noticed that even people who claim that everything is predestined and that we can do nothing to change it look before they cross the road.
>
> Stephen Hawking[1]

Opposition to the idea of free will has existed from ancient times to the present. This chapter begins by discussing doctrines of theological, philosophical, and scientific predeterminism. It then addresses genetic, environmental, and other ad hoc arguments against free will. It concludes with a discussion of what is called "compatibilism."

THEOLOGICAL ARGUMENTS AGAINST FREE WILL

"On the traditional theological determinist conception, God creates us and our environments in such a way that our entire life-histories are intentionally deterministically programmed by him from the beginning of our lives."[2] Although various philosophical and religious systems have followed different forms of theological predeterminism, I am most familiar with Christian versions, and the following discussion is limited to Christian applications of this doctrine.[3]

Paul's Letter to the Romans

Paul's letter to the Romans in the Christian New Testament is the most important source of Christian predeterminism or, as it is often called, predestinarianism. The key text is Romans 8:29–30: "For those whom he foreknew [God] also predestined to be conformed to the image of his Son [Jesus], in order that he might be the firstborn within a large family. And those whom he predestined he also called; and those whom he called he also justified; and those whom he justified he also glorified."[4] Shortly thereafter, Paul speaks of "God's elect."[5] See also Romans 9:11–18, 11:5–6, and Ephesians 2:8–10.

Augustine (354–430)

Augustine of Hippo, later known as St. Augustine, was not born and raised a Christian. Instead, after an extended period of philosophical study, he converted to Christianity and was baptized in that faith in 387. During the years 387–395, he successively wrote the three books of *On Free Choice of the Will*.[6] The first book, written 387–89, reveals a strong Neoplatonic influence. The second and third books, written about 391–95, show an Augustine who is becoming more and more influenced by Christian scripture and teachings.[7] Portions of this work appear to set forth a doctrine of free will. In fact, Pelagius (ca. 360–418), who supported a concept of free will in a Christian context, later used this work against Augustine in a major controversy over free will. By that time, Augustine's views on free will had changed; he now rejected the idea of independent human free will. In the event, Augustine won his battle with Pelagius and succeeded in having the latter declared a heretic.[8]

In 426–27, shortly before the end of his life, Augustine wrote a work called *Reconsiderations*. In book I, chapter 9, he reinterpreted his youthful work *On the Free Choice of the Will* as ignoring the question of God's grace, because that publication was directed against the Manicheans, not the Pelagians. The Pelagian controversy did not occur until much later. In *Reconsiderations*, Augustine argued that the ability to have virtue (the correct exercise of will) depends on God's prior grace, which is extended only to the elect few.[9] After quoting certain passages from *On the Free Choice of the Will*, he commented:

> In these and similar passages I did not mention the grace of God, which was not then under discussion. Consequently, the Pelagians think, or could think, that I held their view. Far from it. As I emphasized in these passages, it is indeed by the will that we sin or live rightly. But unless the will is liberated by grace from its bondage to sin and is helped to overcome its vices, mortals cannot lead pious and righteous lives. And unless the divine grace by which the will is freed preceded the act of the will, it would not be grace at all. It would be given in accordance with the will's merits, whereas grace is given freely. I have dealt satisfactorily with these questions in other

works, refuting these upstart heretics who are the enemies of grace.[10]

One of the writings in which Augustine elaborated his final views on free will is his *Treatise on the Predestination of the Saints* (ca. 428).[11] In chapter 7 of this work, he admits that "I was in . . . error, thinking that faith whereby we believe on God is not God's gift, but that it is in us from ourselves, and that by it we obtain the gifts of God, whereby we may live temperately and righteously and piously in this world."[12] In chapter 16, he states: "Faith, then, as well in its beginning as in its completion, is God's gift; and let no one have any doubt whatever, unless he desires to resist the plainest sacred writings, that this gift is given to some, while to some it is not given."[13] We should not find fault with God for this, he says, because all humans deserve condemnation as a result of original sin (Adam's fall).[14]

In chapter 19, Augustine states that salvation is not by human will but rather "by divine grace or predestination. Further, between grace and predestination there is only this difference, that predestination is the preparation for grace, while grace is the donation itself."[15] In chapter 34, he writes that the elect "were already chosen [by God] before the foundation of the world."[16] "[T]hey were elected before the foundation of the world with that predestination in which God foreknew what He Himself would do; but they were elected out of the world with that calling whereby God fulfilled that which He predestinated."[17] Therefore, he says in the next chapter, "God chose us in Christ before the foundation of the world, predestinating us to the adoption of children, not because we were going to be of ourselves holy and immaculate, but He chose and predestinated us that we might be so."[18]

As we will now see, the leaders of the sixteenth-century Protestant Reformation picked up Augustine's themes to deny the existence of human free will.

Martin Luther (1483–1546)

Martin Luther, the founder of Protestantism, was much influenced by Augustine's predestinarian doctrine. Luther rejected free will in *The Bondage of the Will*[19] and other writings. In his "Preface to Romans," Luther wrote: "In [Romans], Paul deals with the eternal providence of God. It is by this providence that it was first decided who should, and who should not, have faith . . . . This is a matter which is taken out of our hands, and is solely at God's disposal . . . ."[20] Although Luther is famous for his doctrine of salvation by faith and not by works, he, like Augustine, held that faith is a matter of God's grace. Accordingly, God predetermines the destiny of every person.

Luther appeared to go beyond predestinarianism (the question of Christian salvation) to theological predeterminism/determinism in all matters. In *The Bondage of the Will*, Luther wrote that God "foresees, purposes, and does **all things** according to His own immutable, eternal and infallible will."[21] "God foreknows and wills **all things** . . . ."[22]

John Calvin (1509–64)

John Calvin defined predestination as "the eternal decree of God, by which he determined with himself whatever he wished to happen with regard to every man. All are not created on equal terms, but some are preordained to eternal life, others to eternal damnation . . . ." Accordingly, "as each has been created for one or other of these ends, we say that he has been predestined to life or death."[23] Calvin specifically discussed the relationship between predestination and the Reformation doctrine of "justification by faith": "While the elect receive the grace of adoption by faith, their election does not depend on faith but is prior in time and order. As the beginning of faith and perseverance in it arises from the gratuitous election of God, none are truly illuminated with faith, and none granted the spirit of regeneration, except those whom God elects. . . . While we are elected in Christ, nevertheless that God reckons us among his own is prior in order to his making us members of Christ."[24] Calvin's doctrine was not limited to matters of salvation. He wrote that "all events take place by [God's] sovereign appointment."[25]

Thomas Hobbes's Divine Predeterminism

Thomas Hobbes (1588–1679) taught that power is the ultimate fact in the universe—from the absolute power of God to the absolute power of the political sovereign. He elaborated his political philosophy in *Leviathan, or the Matter, Forme, and Power of a Common-Wealth Ecclesiastical and Civill* (1651). There Hobbes argued that the political sovereign (preferably an absolute hereditary monarch) is God's representative on earth and, as such, has divine authority to decide all things, including but not limited to the ecclesiastical form, doctrines, and practices of a national religion to be imposed on all subjects.²⁶ As Leo Strauss observed, "Hobbes's personal attitude toward positive religion was at all times the same: religion must serve the State and is to be esteemed or despised according to the services or disservices rendered to the State."²⁷ Although Hobbes wrote that natural theology could prove nothing about God other than that there was a first cause, he developed "a pretended revealed theology" to support his project of political absolutism.²⁸

Hobbes also used his "pretended revealed theology" to support a doctrine of divine predeterminism. An important source of Hobbes's views on predeterminism involves an extended disputation—first oral, and then written—between Hobbes and John Bramhall, a bishop in the Church of England. In 1645, Hobbes, Bramhall, and the Marquess of Newcastle were all living in exile in France as a result of the English civil war raging at that time. All three were English supporters of the king and the established Anglican church as against the forces of Parliament. During the civil wars, Parliament and later the Cromwell Protectorate were successful in controlling the government for many years before the restoration of the Stuart monarchy in 1660.

The Hobbes-Bramhall disputation began in 1645 at Newcastle's house in Paris. Following that oral debate, Hobbes and Bramhall exchanged written arguments, which they also shared with Newcastle. These written disputations eventually made their way into print, and substantial portions of them were made generally available in the late 1990s.²⁹

In his introduction to this work, Vere Chappell observed that "Hobbes's view of freedom and necessity was quite similar to that of the Protestant Reformers, Luther and Calvin among others."[30] Hobbes explicitly admitted this.[31] As discussed in earlier subsections of the present chapter, Luther and Calvin posited the absolute sovereignty of God. Luther stated that God "foresees, purposes, and does all things according to His own immutable, eternal and infallible will."[32] Calvin likewise wrote that "all events take place by [God's] sovereign appointment."[33]

Hobbes had a similar argument, albeit without all the theological trappings of Luther and Calvin. Hobbes argued that the sum of contributing causes to every action "necessitates and determines" the action and that such **"concourse of causes ... may well be called (in respect they were all set and ordered by the eternal cause of all things, God Almighty) the decree of God.**"[34] Both a person's choice ("election") and the effects of that choice are predetermined ("necessitated").[35] "[E]very volition or act of the will and purpose of" human beings has "a sufficient and therefore a necessary cause; and **consequently every action was necessitated** [predetermined]."[36] "[T]here is no such thing as freedom from necessity [predeterminism]."[37] The denial of necessity (predeterminism) "destroys both **the decrees and the prescience of God Almighty**."[38] "**[T]he will of God makes the necessity of all things.**"[39]

Hobbes elaborated at some length the scriptural justifications for his position, including but not limited to Paul's letter to the Romans, discussed in an earlier subsection of this chapter.[40] Hobbes wrote that "Saint Paul, [who] disputes [argues] that question largely and purposely, never uses the term of 'free-will'; nor did he hold any doctrine equivalent to that which is now called the doctrine of free-will, but derives all actions from the irresistible will of God, and nothing from the will of him that runs or wills."[41]

The crux of Hobbes's position is his assertion that "**all actions**, even of **free** and **voluntary** agents, are **necessary** [predetermined]."[42] This appears to be a contradiction in terms, but it is the essence of what was later known by the name "compatibilism." The contradiction is resolved, according to the compatibilist argument, by redefining "voluntary" and "free" to be compatible with mechanistic predeterminism.[43] Compatibilism is discussed in the final section of this chapter.

Evaluation of Christian Theological Arguments against Free Will

There is a basic continuity in the final views of Paul, Augustine, Luther, and Calvin on free will. All of them adopted a kind of theological predeterminism or predestinarianism that precluded independent human free will. They taught that Christian morality and salvation are made possible only by faith, but faith, in turn, is entirely dependent on God's preexisting grace. God decides, before each person is born, whether that person will be chosen to be among the elect few. Those whom God did not freely choose are predetermined to be incapable of true Christian virtue and to be consigned, after death, to eternal damnation. Luther and Calvin went so far as to argue that everything—not just matters of faith and salvation—is predetermined. Hobbes picked up the latter point and made it central to his doctrine of divine predeterminism.

Christian predestinarianism is based on alleged divine revelation. It is not for everyone to be able to have faith in such revelation; that depends—according to Augustine, Luther, and Calvin—on God's prior grace. We are aware in the modern age that there are many mutually exclusive claims to revelation.[44] It is impossible to choose among the competing claims to revelation on the basis of reason. Indeed, according to these famous Christian theologians, reason has nothing to do with it. Rather, it is entirely a matter of faith, and faith is solely the result of God's preexisting election. If all of the foregoing is true, it is impossible to attempt to formulate a rational ethical philosophy. God has already done so for us, but only the elect will follow it in any case. Any human effort to develop a rational ethics is, in effect, blasphemous.

In his *Lectures on the Philosophical Doctrine of Religion*, the philosopher Immanuel Kant stated the following about the doctrine of divine predestination:

> [This doctrine] is absolutely improper regarding God; for such a thing would make of God not only a despot but a complete tyrant, as if without any regard to the worthiness of his subjects he elected some to happiness and condemned the others straightway to reprobation, providing all sorts of remedies for the first and withdrawing from the others every power and opportunity to make themselves worthy of happiness, so as to do all this with a show of right. . . .

> Insofar as its object is the reprobation of one whole part of humankind, the doctrine of *predestination* presupposes an *immoral* order of nature. For it is thereby asserted that in the case of some human beings the circumstances of their lives are so ordered and connected that they could not but be unworthy of blessedness.[45]

Although I disagree with Kant on some issues, I agree with his foregoing statement. Chapter 2 discusses Kant's views on free will in further depth.

Interestingly, much of popular Christianity later deviated from the stern predestinarianism of Paul, Augustine, Luther, and Calvin. Traditional Christianity never accepted the unorthodox and bizarre scriptural arguments of Hobbes, who was considered an infidel. The next chapter discusses, among other things, the development of some limited notions of free will in recent Christianity.

PHILOSOPHICAL AND SCIENTIFIC ARGUMENTS AGAINST FREE WILL

As astrophysicist-philosopher Bob (Robert O.) Doyle has observed, "There is little in philosophy more dehumanizing than the logic chopping and sophisticated linguistic analysis that denies the possibility of human freedom."[46] The present book will spare the reader the technical minutia and jargon of the contemporary academic attack on free will. Those interested in exploring the voluminous scholarly literature on this subject can consult the references cited in the endnotes and bibliography. The following discussion summarizes and evaluates the main positions.

Pierre-Simon Laplace (1749–1827)

The trope of Pierre-Simon Laplace's "demon" has long been the most famous statement of the principle of predeterminism in classical physics:

> All events, even those which on account of their insignificance do not seem to follow the great laws of nature, are a result of it just as necessarily as the revolutions of the sun. In ignorance of the ties which unite such events to the entire system of the universe, they have been made to depend upon final causes or upon hazard, according as they occur and are repeated with regularity, or appear without regard to order; but these imaginary causes have gradually receded with the widening bounds of knowledge and disappear entirely before sound philosophy, which sees in them only the expression of our ignorance of the true causes.
>
> Present events are connected with preceding ones by a tie based upon the evident principle that a thing cannot occur without a cause which produces it. . . .
>
> We ought then to regard the present state of the universe as the effect of its anterior state and as the cause of the one which is to follow. Given for one instant an intelligence which could comprehend all the forces by which nature is animated and the respective

> situation of the beings who compose it—an intelligence sufficiently vast to submit these data to analysis—it would embrace in the same formula the movements of the greatest bodies of the universe and those of the lightest atom; for it, nothing would be uncertain and the future, as the past, would be present to its eyes.[47]

This predeterministic view of one possible future dominated classical physics from the time of Newton to the discovery of quantum mechanics in the twentieth century. Christian List has recently summarized this mindset as follows:

> The thesis that the world is deterministic is often considered a hallmark of a scientific worldview. Determinism implies that, given the initial state of the universe, only one course of events will have been physically possible. If the world is deterministic, there could never be any alternative possibilities. The initial state of the world—say, at the time of the Big Bang—together with the laws of nature, would have been sufficient to determine all subsequent events, from the motion of the planets to all human behaviour. Your reading this book now would be nothing but an inevitable consequence of the physical past. You would never have had any genuine choices at all.[48]

Needless to say, this is a very bleak picture. It is difficult for most people to grasp how anyone could assert such a proposition. Yet, as we will now see, this view, or some variation of it, is widely accepted in today's academic world.

Ted Honderich

Ted Honderich, a British philosophy professor, is the leading proponent in our time of the doctrine known as determinism (more accurately called "predeterminism"). His most elaborate treatment of this issue was in his 1988 tome *A Theory of Determinism: The Mind, Neuroscience, and Life*. But, as he explains in the second edition (2002) of his book *How Free Are You?: The Determinism Problem*, "I've had some second thoughts," and this second edition "goes further than its large predecessor and also than the first edition of itself."[49] Accordingly, the present discussion focuses on the second edition of *How Free Are You?*, and the page numbers cited in textual parentheses refer to that edition. The endnotes additionally reference some of his later essays on this subject.

Honderich defines "determinism" as "a theory that all our mental events, including choices and decisions, and also our actions, are effects of certain things and therefore have to happen or are necessitated, and cannot be owed to origination" (155).[50] He defines "free will" as "a kind or part of freedom that is or rests on our supposed personal power to originate choices and thus actions—i.e. origination as a power; sometimes used more generally" (156). Accordingly, if determinism, as defined by Honderich, exists, free will does not. Indeed, his premise is that free will does not exist.

Honderich concludes that "taking in account everything, determinism is *very strongly supported*, and that certainly it has *not been shown to be false*. It will be no news to you that I myself *do* think determinism is true, but that thought does seem even to me to go a bit beyond the evidence" (90 [italics in the original]). "Furthermore, ... it does not matter if determinism is true or false, since there is no serious idea with which it conflicts. The question of its truth does not need looking into. That was time wasted" (103 [citation omitted]).

This is an astonishing position for a philosophy professor to take. He feels no need to demonstrate the truth of (pre)determinism, because the advocates of free will cannot, in his view, absolutely and scientifically prove the case for free will. As with so many (pre)determinist opponents of free will, he attempts to place the burden of proof on the advocates of free will, notwithstanding the fact that a belief in some kind of free will is consistent with human experience whereas a belief in (pre)determinism with regard to human choices and decisions is counterintuitive.

Honderich teaches that all "events" (including all mental operations) are strictly a matter of physical cause and effect. Ergo, all our thoughts are caused by other physical events—ultimately events outside of ourselves going back to the beginning of time. Everything is predetermined. (Pre)determinism is "the question of whether your choosing this book and your reading this sentence now, or your deciding to move in with someone or get divorced, is just a matter of cause and effect" (1). He believes that the answer to this question is in the affirmative.

Honderich is aware that quantum physics claims to have discovered a world governed by indeterministic rather than deterministic principles, but he denies that such indeterminism is ontologically real (74–75).[51] In any event, he argues, like many predeterminists or quasi-predeterminists, that any such phenomena would be of no account at the level of human choices and decisions:

> The same question of consequences, by the way, is raised by something perhaps more widely accepted than determinism. That is *near-determinism*. Maybe it should have been called *determinism-where-it-matters*. It allows that there is or may be some indeterminism but only at what is called the micro-level of our existence, the level of the small particles of our bodies, particles of the kind studied by physics. At the ordinary level of choices and actions, and even ordinary electrochemical activity in our brains, causal laws govern what happens. It's all cause and effect in what you might call real life. (5 [italics in the original])

Honderich's view that quantum mechanics would (if it were real) apply only to the microworld is a standard trope among predeterminists or quasi-predeterminists. In contrast, quantum physicist Henry Stapp writes: "The oft-heard claim that 'quantum mechanics is not relevant to the mind-brain problem because quantum theory is only about tiny things', is absolutely contrary to the basic quantum principles."[52] Stapp's arguments in support of free will are discussed in Chapter 2.

Additionally, Honderich and other predeterminists wrote their major works before the publication of relevant books by neuroscientists William R. (W. R.)

Klemm and Peter Ulric Tse, research psychiatrists Jeffrey M. Schwartz and Norman Doidge, and astrophysicist-philosopher Robert O. (Bob) Doyle. As discussed in Chapter 2, the writings of such scientifically well-informed authors present significant alternatives to the predeterminist assumptions of Honderich and his intellectual kin—alternatives that are based on science, evidence, and reason.

Chapter 11 of Honderich's book evinces, among other things, his animus against retribution as a basis for criminal punishment. In this and other writings, he is passionately opposed to any notion of retribution as a legitimate consideration in criminal or social matters. To consign retribution to oblivion, he proposes to replace any notion of free will with the doctrine of predeterminism: if the criminal was predetermined to commit the crime, he/she cannot be responsible for it. This is the equivalent of killing a fly with a hydrogen bomb. There are several approaches to criminal punishment, for example, retribution, deterrence, societal protection, and rehabilitation. Some criminal law theorists oppose retribution as a basis for criminal law without any reference to (pre)determinism.[53] Although good arguments can be made against retribution, it must also be kept in mind that the social contract involves the assumption by government of the formerly private right of retribution/retaliation. See John Locke's *Second Treatise of Government*. In any event, Honderich's project of eliminating retribution by means of indoctrination in predeterminist lore is a fool's errand. People are less likely to accept predeterminism than to rethink the wisdom *vel non* of retribution. Moreover, if predeterminism is true, both crime and punishment are inevitable: it would be all a matter of cause and effect.

I tend to agree with the position that, as a matter of political and legal philosophy, the objectives of deterrence, societal protection, and rehabilitation should supersede retribution as the basis of criminal law punishment theory. But, unlike Honderich, I do not base this principle on a premise that people lack all free will. Accordingly, I will elaborate my views on this question in my forthcoming work on political philosophy and not in the present book on free will.

The Libet and Successor Experiments

Benjamin Libet (1916–2007) was a physiologist and experimental neuroscientist. Many predeterminists argue that certain experiments conducted by Libet and his successors constitute proof positive of predeterminism and the nonexistence of free will.[54]

Libet summarized his experiments regarding free will in Chapter 4 of his book *Mind Time: The Temporal Factor in Consciousness*.[55] The parenthetical page numbers in the present subsection refer to this book, published in 2004 (about three years before his death).

Libet's "experimental question was: Does the conscious will to act precede or follow the brain's action?" (130). He concluded from his experiments that "the brain exhibited an initiating process, beginning 550 msec [milliseconds] before the freely voluntary act; but the awareness of the conscious will to perform the act appeared only 150-200 msec before the act. The voluntary process is therefore initiated unconsciously, some 400 msec before the subject becomes aware of her will or intention to perform the act" (123–24).

Libet's experimental design was as follows. He assumed that an electrical charge, called the "readiness potential" (RP) and recorded by "suitable electrodes" on a human subject's head, accurately measured the brain's initiation of a voluntary act (126–27). The subject "was asked to perform a freely voluntary act, a simple but sudden flexion of the wrist at any time he felt like doing so" (126). The subject had only six seconds to flex his wrist (130). The subject was seated about 2.3 meters from a cathode ray oscilloscope that "was arranged to have its spot of light revolve near the outer edge of its face. The outer edge of the oscilloscope tube face was marked in clock seconds, sixty as usual, around the circle. The movement of the light spot was designed to simulate the sweep of the second hand of a usual clock. But our light spot completed the circle in 2.56 sec, about twenty-five times faster than the normal 60 sec . . . . This faster 'clock' could then reveal time differences in hundreds of milliseconds" (126–27).

The subject would "report the 'clock time' for her experience of the conscious intention to act. The clock time would be noted silently and reported after each trial was over" (126). Thus, the subject would not report the clock time for her "conscious intention to act" until after she had flexed her wrist.

These experiments resulted in Libet's conclusion that the brain initiates the flexion movement before the subject consciously intends to act. Although Libet did not undertake any further research on this issue, he generalized the results of his study to speculate that the same result would be found in more complicated human situations (148). Nevertheless, he admitted that he had no experimental evidence relevant to deliberation or the advance making of choices, except for a minor deviation occurring when some of his subjects reported that they had, contrary to instructions, planned when they would flex their wrists during the six seconds allotted to them (132, 148–49).

Contrary to conventional wisdom, these experiments conducted by Libet and his associates do not prove predeterminism and do not disprove free will.

First, there are many technical problems that scientific and other experts have observed with regard to Libet's experimental protocol.[56] For example:

- The timing of when the participants thought they made their decisions was dependent on their scientifically unreliable introspective and subjective reports.[57]
- Only a small part of the brain was monitored for when a decision was actually made, but other portions of the brain, not considered, are likely implicated in decisionmaking, as indicated by other studies.[58] "Nobody knows where in the brain the conscious self is, much less where intentions are first initiated."[59] "[C]omplex cognitive processes cannot be explained by simplistic experiments, no matter how ingenious . . . ."[60]
- Scientific evidence does not establish that what Libet called the "readiness potential" constitutes the formation of an intention.[61] Later scientific experiments demonstrated that Libet's readiness potential was "caused primarily—if not exclusively—by the requirements to monitor the clock display

and to report the spot's position at the moment of the spontaneous decision to move."[62] Neuroscientist Peter Ulric Tse observed in 2013: "Despite three decades of philosophizing about Libet's results, the precise roles of components of the readiness potential have not been empirically established. It is still unclear whether the readiness potential is a neural correlate of the motor act, the planning of the motor act, expectation of a motor act, or the act of consciously willing."[63] Quantum physicist Henry Stapp has argued that "Libet, mistakenly from th[e] quantum point of view, associated the rise of the readiness potential with a decision to act.... But, according to the quantum model, the early part of this rise is merely a concomitant of the process of constructing a 'template for action' that will *only later, by virtue of a mental choice,* be picked out from among the many potential templates that have been constructed in parallel by the Schrödinger-equation-controlled evolution of the quantum mechanical state of the brain of an observer."[64]
- As a result of the foregoing and other considerations, Libet's experiments do not prove the absence of free will in the initiation of actions.[65]

Second, the simple motor experiments of Libet do not involve the frequent instances of extended human deliberation.[66] "Willing a stereotyped, well-rehearsed finger movement is too simple to have much bearing on such conscious processes as the decisions made through planning a course of action that spans past and future, or analysis of complex events. Why, therefore, would anybody be surprised at absence of a robust antecedent indicator of willed finger movement?"[67] "What experiment could cast doubt on the free will involved in self talk, setting goals, making plans, adjusting attitude, developing belief systems, or any decisions or choice not involving action or active refusal to act?"[68]

The assumption, without further study, that brain operations involving simple motor acts operate in the same way as complicated human thought and deliberation exhibits the fallacy of hasty generalization, otherwise known as jumping to conclusions, as well as the fallacy of reductionism.[69]

Finally, Libet himself pointed to his experimental result showing that the conscious will appears 150 milliseconds before the performance of a motor act. Although he assumed that the volitional process began unconsciously with the even earlier "readiness potential," he surmised that the conscious will effectively had 100 milliseconds to "block or 'veto' the process, so that no motor act occurs" (137–38). When the subject preplanned the execution of an act (in violation of instructions), that decisional interval was extended to 100–200 milliseconds (138–38).

Libet observed that "[v]etoing of an urge to act is a common experience for all of us. It occurs especially when the projected act is regarded as socially unacceptable, or not in in accord with one's overall personality or values" (138). He made a scientific argument for the proposition that such conscious veto does not have a preceding unconscious origin (145–47, 202). In his view, "the crucial point is that we have *conscious control* over the actual performance of our unconsciously initiated volitional processes. Hence, we are responsible for our conscious control choices, not for our unconsciously initiated urges that precede our conscious decisions" (208 [italics in the original]).

Libet added: "We may view voluntary acts as beginning with unconscious initiatives being 'burbled up' by the brain. The conscious will would then select which of these initiatives may go forward to an action, or which ones to veto and abort so no motor act appears" (139). This understanding is similar to the conceptions of free will articulated by astrophysicist-philosopher Robert O. (Bob) Doyle and quantum physicist Henry P. Stapp, both of whom are discussed in the next chapter.[70]

Libet explicitly rejected (pre)determinism, noting that "natural [physical] laws were derived from observations of physical objects, not from subjective mental phenomena" (152–53). "I can say categorically that there is nothing in neuroscience or in modern physics that compels us to accept the theories of determinism and reductionism" (216). "[W]e do not have a scientific answer to the question of which theory (determinism or nondeterminism) correctly describes the nature of free will" (154). "Great care should be taken not to believe allegedly scientific conclusions about our nature that depend on hidden ad hoc assumptions" (155). The existence of free will "is at least as good, if not a better, scientific option than is its denial by natural law determinist theory. Given the speculative nature of both determinist and nondeterminist theories, why not adopt the view that we do have free will (until some real contradictory evidence appears, if it ever does)? Such a view would at least allow us to proceed in a way that accepts and accommodates our own deep feeling that we do have free will. We would not need to view ourselves as machines that act in a manner completely controlled by known physical laws" (156).

Several post-Libet experiments have been conducted by others. We cannot here, without engaging in lengthy technical analysis beyond the scope of the present book, discuss this research in depth. However, the interested reader can consult the references in the following endnote for the defects that experts have found with these studies. Many of these problems are similar to those discussed above with regard to the Libet experiments.[71]

Daniel M. Wegner (1948–2013)

One of the most famous books in the free will literature is Daniel M. Wegner's *The Illusion of Conscious Will*.[72] In this work, Wegner claimed that conscious will is an illusion. He argued that the real action in all human causation occurs unconsciously. Ergo, free will, like conscious will, is an illusion.

Wegner stated his essential thesis with remarkable clarity in the last paragraph of chapter 3 of this book:

> The unique human convenience of conscious thoughts that preview our actions gives us the privilege of feeling we willfully cause what we do. **In fact, however, unconscious and inscrutable mechanisms create** both conscious thought about action

> and **the action**, and also produce the sense of will we experience by perceiving the thought as cause of the action. So, while our thoughts may have deep, important, and unconscious causal connections to our actions, the experience of conscious will arises from a process that interprets these connections, not from the connections themselves. (emphasis added)

Accordingly, in chapter 9 of his book, Wegner postulated that free will is a mere emotion, an illusion.

Wegner relied heavily on the Libet experiments that purportedly showed that conscious will does not initiate simple motor actions.[73] We have exposed above the defects in Libet's experimental protocol and philosophical analysis that led to his erroneous view that unconscious processes, not conscious will, initiate action. Wegner did not even mention Libet's further argument that humans have a veto power that enables free will. In fact, Libet himself explicitly rejected Wegner's arguments:

> Given that the issue is so fundamentally important to our view of who we are, a claim that our free will is illusory should be based on fairly direct evidence. Theories are supposed to explain observations, not do away with them or distort them, unless there is powerful evidence to justify that. Such evidence is not available, and determinists have not proposed any potential experimental design to test their theory. The elaborate proposals that free will is illusory,

like that of Wegner [citation omitted] fall into this category. It is foolish to give up our view of ourselves as having some freedom of action and of not being predetermined robots on the basis of an unproven theory of determinism.[74]

Apart from Wegner's misapplication of the Libet studies, he assumed that if some illusory experiences about human causation occur, then all beliefs about conscious human causation are false. He cited alien hand syndrome, phantom limbs, table turning seances, schizophrenia, hypnosis, and so forth.[75] This is the classic fallacy of hasty generalization or jumping to conclusions, discussed above in a somewhat different context with regard to Libet. The instances of illusory causation adduced by Wegner do **not** prove that **all** human experiences of causation are illusory.[76]

Wegner also argued that certain automatic behavior constitutes evidence of unconscious causation.[77] For example, he stated: "Playing tennis or reading or typing or walking are all skills that involve exceedingly fast reactions, many automatic in this sense."[78] As the Introduction of the present book explained, however, mastery of such activities involves a training sequence in which a person has to learn, in a very conscious way, the step-by-step (literally, in the case of walking) process of learning to do them. After much practice, these activities do, indeed, become a matter of partially unconscious habit (see also the discussion of neuroplasticity in Chapters 2 and 3). But Wegner wrote as though the earlier slow, conscious learning process never occurred and that these activities suddenly sprang from the unconscious brain like Athena from the brow of Zeus.

Similarly, Wegner adduced examples of people experiencing what he called "creative insight" when they are not even thinking about the subject in question. For example, a solution to a mathematical problem unexpectedly occurs to a mathematician on vacation. Einstein reported having suddenly understood a way toward resolution of a physics issue while working on something else in his day job at the patent office. One could adduce many other examples, for example grasping the solution to a practical or theoretical problem when awakening from sleep. Wegner suggested that such great moments of insight derive from the unconscious instead of from conscious cogitation and will. Again, he ignored the fact that all such "inspiration" is prepared by long, conscious preparation in the subject matter. $E = mc^2$ does not pop into the head of a person who lacks an in-depth training in mathematics and physics. Someone without Mozart's extensive background in music (starting at age three)[79] could not have composed Mozart's great works, some of which came to his mind while attending social functions. Yes, the unconscious plays a role in creative insight but not without a previous conscious and often laborious training and experience in the field in which the unexpected thought occurs. As the popular saying goes, genius is 99% perspiration and 1% inspiration (probably in such cases as Mozart and Einstein the inspiration—not to mention the innate ability—is a greater percentage). Intent on "proving" that the unconscious rules all human thought and action,

Wegner ignored facts that are obvious to every impartial observer.[80]

It is astonishing that a person of Wegner's professional stature would commit such elementary logical fallacies—and even more amazing that so many intellectuals of our time would uncritically accept them as gospel.[81] As an eminent political philosopher once wrote in another context, such blunders would shame an intelligent high school student.[82]

Sam Harris

In his popular book *Free Will*, neuroscientist Sam Harris alleges that "[f]ree will *is* an illusion. Our wills are simply not of our own making. Thoughts and intentions emerge from background causes of which we are unaware and over which we exert no conscious control. We do not have the freedom we think we have."[83]

Similarly, Harris opines:

> Choices, efforts, intentions, and reasoning influence our behavior—but they are themselves part of a chain of causes that precede conscious awareness and over which we exert no ultimate control. My choices matter—and there are paths toward making wiser ones—but I cannot choose what I choose. And if it ever appears that I do—for instance, after going back and forth between two options—I do not *choose* to choose what I choose.[84]

In the course of reaching such conclusions, Harris cites the studies of Benjamin Libet and Daniel Wegner.

The preceding two subsections of this book demonstrated that the arguments of Libet and Wegner about the allegedly unconscious initiation of human action are based on empirical missteps and logical fallacies. Accordingly, their positions on these matters are scientifically and philosophically untenable.

Although Harris is a self-professed atheist, he prosecutes his case against free will with religious zeal. Except for the debunked analyses of Libet and Wegner, he offers little in the way of evidence to support his articles of faith. His case against free will depends on unexamined premises rather than on reason and evidence.

Susan Blackmore

Susan Blackmore is a psychologist, writer, and professor. Her essay "Living without Free Will,"[85] is based on the premise that free will is merely a popular illusion that does not, in fact, exist. The only question for her is how one can live without this illusion. She notes two possible solutions: "One is to go on living 'as if' we have free will—in other words, to accept that free will is an illusion and yet choose to remain deluded (not a free choice of course, but one caused by prior events and circumstances). The other is to reject the illusion and aspire to live entirely without free will."[86]

She concludes, on the basis of personal interviews, that such famous free will skeptics as Daniel Dennett and Daniel Wegner choose to live "as if" free will exists while actually believing that it does not. In contrast, she herself attempts to live her life with full, ever-present knowledge that free will does not exist. The latter course of action leads her into what she understands to be Buddhist thought.

Blackmore's assertion that the choice of one of these two alternatives is "not a free choice . . . but one caused by prior events and circumstances" evinces her commitment to predeterminism even when it defies common sense. She is less consistent, however, in her approach to criminal justice. One of her colleagues stated, "if no one has free will, it means that no one should be in prison . . . [H]ow can it provide a deterrent for people if they don't have free will; it's not up to them[?]" Blackmore responded as follows:

> In contrast, I think that the criminal justice system would be stronger and fairer if it were not based on the notion of free will. Certainly we would lose the idea of retribution; of punishing people because they acted badly of their own free will and so deserve to suffer. But people would be sent to prison for other reasons: to keep them away from doing any more harm, for training or rehabilitation, or as a deterrent to them or others in the future. We know that appropriate rewards and punishments can change people's behavior. So the relevant question would not be "does this person deserve to be punished?" but "would this punishment do any good to them, to their victims, or to society in general?" In many cases, the answer would be "yes."[87]

As with Ted Honderich (see above), Blackmore is obsessed with the idea of abolishing retribution as a basis of criminal justice. One wonders from such passion whether the antiretributionist tail is wagging the predeterminist dog. But the issue of whether retribution is properly a basis for criminal justice has long been a question of political and legal philosophy without any consideration of the truth *vel non* of predeterminism. As discussed above, it would be much more difficult to persuade most people that free will is an illusion than to show them that deterrence, societal protection, and rehabilitation are sufficient bases of criminal justice without any need for retributionist furor. Note, moreover, that Blackmore failed to answer her colleague's question about how prison would be a deterrent if free will did not exist. If one takes the predeterminist philosophy to its logical conclusion, nothing we do (all of which is predetermined) can deter crime, which is inevitable.

GENETIC, ENVIRONMENTAL, AND OTHER AD HOC ARGUMENTS AGAINST FREE WILL

F. Scott Fitzgerald opened his famous novel *The Great Gatsby* with the following lines in the voice of his fictional first-person narrator, Nick Carraway:

> In my younger and more vulnerable years my father gave me some advice that I've been turning over in my mind ever since.
>
> "Whenever you feel like criticizing anyone," he told me, "just remember that all the people in the world haven't had the advantages that you've had."
>
> He didn't say more but we've always been unusually communicative in a reserved

> way and I understood that he meant a great deal more than that. In consequence I'm inclined to reserve all judgements, a habit that has opened up many curious natures to me and also made me the victim of not a few veteran bores . . . . Reserving judgements is a matter of infinite hope. I am still a little afraid of missing something if I forget that, as my father snobbishly suggested and I snobbishly repeat, a sense of the fundamental decencies is parceled out unequally at birth.
>
> And, after boasting this way of my tolerance, I come to the admission that it has a limit.[88]

Reserving moral judgment—indeed, the alleged invalidity of all moral judgments—is a theme of many recent attacks on free will. To quote just one of many contemporary scholars, "agents are never morally responsible for their actions."[89] It is ironic—mostly likely a deliberate irony—that Fitzgerald put certain predeterministic statements in the mouth of Nick Carraway, since *The Great Gatsby* contains page after page of explicit and implicit moral evaluations and judgments—most of them expressed, however indirectly, by Carraway himself. I have always considered *Gatsby* to be a masterpiece of ethical and political commentary, though some of its statements and characters are below our present standards of political correctness. With regard to the latter, one must observe that it was written during the 1920s—long before our current awareness of the importance of certain values. That does not excuse its author; it merely points to some of his limitations. Otherwise, his novel is a work of genius.

One cannot argue with the proposition that genetic and environmental factors sometimes—perhaps even often—influence human thought and behavior. However, as neuroscientist William R. (W. R.) Klemm has stated, "Intentions, decisions, and choices are of course influenced by their unconscious antecedents, but are not inevitably determined by them because the conscious mind can intervene, veto, or otherwise control."[90]

In light of the scientific findings regarding quantum physics, the "hard determinism" (predeterminism) of yesteryear is gradually giving way to an agnosticism regarding universal determinism (the view that "all events are universally determined by natural causes"[91]). The new view is that "recent developments in the behavioral, cognitive, and neurosciences appear to threaten many of our core presuppositions about freedom and autonomy . . . ."[92] This is a very significant shift in analysis. The proponents of genetic and environmental determinism like to conflate their analysis with the traditional hard determinism of Laplace and his predecessors and successors. But there is a world of difference between old-fashioned determinism and newfangled determinism. The old-fashioned determinism involved predeterminism and inevitability. The new determinism is ad hoc; much depends on individual circumstances. Thus, Caruso, a self-described "hard-enough" determinist, admits that the "recent advances in psychology, sociology, and neuroscience," "unlike the previous family of concerns" (the hard predeterminism of classical physics), "do not presuppose any particular account of physics."⁹³ Christian List, a professor of philosophy and political science, observes:

> Now, I should acknowledge that, in both public and scientific discourse, we keep hearing claims to the effect that certain forms of human behaviour are explained by people's genes, their social and economic backgrounds, their upbringing and education, their peers, and other factors. But those research findings are rarely able to attribute more than a certain *part* of the observed behavioural variation across people to those purported explanatory variables. The variables in question are typically shown, at most, to affect the *probabilities* of certain behaviours, without fully determining them, and there is little reason to think that we will ever arrive at a truly deterministic theory of psychology.⁹⁴

Similarly, neuroscientist Antonio Damasio has remarked:

> [A]lthough biology and culture often determine our reasoning, directly or indirectly, and may seem to limit the exercise of individual freedom, we must recognize that humans do have some room for such freedom, for willing and performing actions that may go against the apparent grain of biology and culture. Some sublime human achievements come from rejecting what biology or culture propels individuals

to do. Such achievements are the affirmation of a new level of being in which one can invent new artifacts and forge more just ways of existing.[95]

A related argument against free will is based on experiments purporting to demonstrate that people are sometimes fooled into thinking that they are making certain conscious decisions when their actions may in fact be motivated by some unconscious (automatic) or subconscious (say, Freudian, to go to the source of this way of thinking) mechanism. As established above, such experiments do **not** establish that **all** human decisions are unconscious or subconscious. Accordingly, they do not prove scientific predeterminism or disprove free will as we have defined it above in the Introduction.[96]

Human experience suggests that individual human beings often have the power to rise above their particular genetic and environmental influences. No, they cannot be taller than the height dictated by their genetic inheritance. But, as Aristotle recognized long ago in his *Nicomachean Ethics*, they can work to improve their ethical and intellectual abilities. Even some contemporary scholars who disagree with popular conceptions of free will do not deny this fact. For example, Thomas W. Clark writes: "In short, although the threat of unconscious factors to integrity and autonomy is very real, conscious processes, once engaged, have the capacity to counteract their influence."[97] Chapters 2 and 3 of this book and my forthcoming book *Reason and Human Ethics* discuss these considerations in greater depth. Indeed, this is what ethics is all about.

COMPATIBILISM AND SOFT DETERMINISM

"Compatibilism" is the view that free will and predeterminism (often called "determinism" or, in earlier times, "necessity") are compatible.[98]

One way of thinking about compatibilism is to acknowledge that mechanistic predeterminism and/or mechanistic indeterminism govern inorganic matter, whereas biological entities possess evolutionary, emergent qualities that result in behavioral free will in lower organisms and more developed free will in mammals and especially in human beings.[99] This is essentially my own view.

However, that is not what most writers calling themselves "compatibilists" mean by compatibilism. The following discussion addresses the more common uses of the term, which has also been labeled "soft determinism."

In the Introduction to the present book, I defined "free will" as "the independent ability to make conscious decisions that are neither predetermined nor random." Compatibilists of the soft determinist variety implicitly or (less often) explicitly reject this definition. Instead, they rely on other (often implicit) definitions of free will and predeterminism. The reader must be careful not to fall into the semantic trap of thinking that all definitions of free will are the same.

As discussed earlier in this chapter, Thomas Hobbes advocated a doctrine of divine predeterminism, which appears to have been a fig leaf for scientific predeterminism. Hobbes nevertheless claimed that his view was compatible with free will, because he defined free will as merely the absence of external coercion. In his convoluted seventeenth-century language, he wrote: "I do indeed take all voluntary acts to be free, and all free acts to be voluntary; but withal that all acts, whether free or voluntary, if they be acts, were necessary [predetermined] before they were acts."[100] Hobbes was able to maintain the alleged compatibility of predeterminism and free will only by redefining free will in terms of external acts rather than (as in my definition) by internal choices and decisions. This is the classic "soft determinist" position.

Immanuel Kant called this kind of compatibilism a "wretched subterfuge" that "would at bottom be nothing better than the freedom of a turnspit, which, when once it is wound up, also accomplishes its movements of itself."[101] William James characterized the Hobbesian approach (which he explicitly called "soft determinism") as "a quagmire of evasion under which the real issue of fact has been entirely smothered."[102]

The philosopher David Hume (1711–76) has also been called a compatibilist or soft determinist along the same line as Hobbes. Although Hume's mature statements on the issue are more ambiguous than those of Hobbes, he may have had similar views, in which case, again, he succeeds only by redefining "free will" out of existence.[103]

In our own time, professional philosopher Daniel C. Dennett is a self-acknowledged "compatibilist"—one who takes a middle road between the "hard determinists" and the advocates of free will. In his book *Freedom Evolves*, Dennett writes that "compatibilism [is] the view that free will and determinism are compatible after all . . . ."[104]

ARGUMENTS AGAINST FREE WILL

How Dennett can take such a position without violating the principle of (non)contradiction is the central mystery of this work. He tries to accomplish it by utilizing semantic legerdemain: changing the historical meanings of such terms as "determinism," "inevitability," and "free will" so that they signify something other than what they have normally meant in the millennia of philosophical, scientific, and other debate on these issues. In Dennett's terminological universe, "determinism" does not always imply what "determinism" is standardly taken to imply, and "free will" does not mean **free will**.[105]

For example, Dennett remarks in chapter 4: "The hard determinists among you may find in subsequent chapters that your *considered* view is that whereas free will—as you understand the term—truly doesn't exist, something *rather like* free will does exist, and it's just what the doctor ordered for shoring up your moral convictions, permitting you to make the distinctions you need to make. Such a soft landing for a hard determinist is perhaps only terminologically different from *compatibilism*, the view that free will and determinism are compatible after all, the view that I am defending in this book."[106] He does not clearly explain what he means by "something *rather like* free will."

Dennett's earlier book *Elbow Room: The Varieties of Free Will Worth Wanting* stated that "*the past does not control us.* . . . Far from it being the case that we are completely under the control of our ancestors or our evolutionary past, it is rather the case that that heritage has tended to set us up as self-controllers—lucky us."[107] But he also argued: "It is often said that no one can change the past. This is true enough, but it is seldom added that no one can change the future either. If the past is unchangeable, the future is unavoidable—on anyone's account."[108] In reaching this bizarre conclusion, Dennett played the kinds of semantic word games that he later employed in *Freedom Evolves*.[109] By this "logic," one is not only powerless to affect the future of one's own life, but political leaders are also unable to influence the future course of political events. Nuclear war, if it occurs, will be, to use Dennett's jargon, "unavoidable" but not "inevitable," just as earthquakes are unavoidable. This dubious semantic distinction would be cold comfort to those, if any, who survived such a catastrophe.

Nevertheless, Dennett somehow concludes in *Elbow Room* that free will "worth wanting" is not an illusion.[110] What, **exactly**, free will "worth wanting" **is** remains unclear. However, Dennett certainly would not agree with my own definition of free will (see the Introduction, above). In his 2015 preface to the new edition of *Elbow Room* (originally published in 1984), he explicitly states that certain concepts (which are implied in my own definition of free will) "don't, and can't, exist, but although some philosophers still take them seriously, they are of only historical interest, like mermaids and leprechauns."[111] In fact, at the time of writing this 2015 Preface, Dennett was exploring the possibility of abandoning the term "free will" altogether, suggesting that he had earlier discussed the concept only in deference to popular opinion.[112]

In Chapter 15 of his 1978 book *Brainstorms*,[113] Dennett proffered a two-stage model of free will that is similar to the model articulated by astrophysicist-philosopher Bob Doyle (see the discussion of Doyle in Chapter 2) and others. In Dennett's words:

> The model of decision making I am proposing has the following feature: when we are faced with an important decision, a consideration-generator whose output is to some degree **undetermined** produces a series of considerations, some of which may of course be immediately rejected as irrelevant by the agent (consciously or unconsciously). Those considerations that are selected by the agent as having a more than negligible bearing on the decision then figure in a reasoning process, and if the agent is in the main reasonable, those considerations ultimately serve as predictors and explicators of the agent's final decision.[114]

But after painting this attractive portrait for advocates of free will, Dennett rejects it. In the end, he claims that the "consideration-generator" must be deterministic, not indeterministic: "Computers are typically equipped with a random number generator, but the process that generates the sequence is a perfectly deterministic and determinate process."[115] Since Dennett does not see any essential difference between computers and brains (he says elsewhere that we are nothing but "moist robots"[116]), he thinks that biological agents must operate in the same deterministic manner as artificial intelligence.[117] He retained that view in the Preface (2017) to the fortieth anniversary edition of *Brainstorms*: "Some version of [the two-stage] model—

minus the indeterminism—is all we need for free will worth wanting in this Godless world."[118]

Accordingly, we finally have Dennett's answer to the question of what "free will worth wanting" is. It is the **appearance** of free will that is worth wanting. The **reality**, in Dennett's clear, cold eyes, is that everything is deterministic. This is just another application of "soft determinism," which Dennett prefers to call "compatibilism."

The rest of the contemporary academic debate over compatibilism/soft determinism has revolved around similar issues, all or most of which involve an implicit or explicit redefinition of free will in order to focus on actions instead of free mental choices and decisions.[119] There is no need to rehearse these arguments here, as I reject the very premise of compatibilism/soft determinism: that free will essentially involves **actions** rather than **thoughts**. The compatibilist/soft determinist arguments shift the entire debate over free will into an obsession with what a person can **do**, as distinguished from what a person can **think, choose, and decide**. This redefinition of free will devolves, in turn, into abstruse academic discussions about moral and legal responsibility with regard to **actions**. Such questions are within the purview of my future works on ethics and political philosophy. They are not relevant to the present book, except to the extent that the logical consequences of hard or soft determinism bear on the truth of those theories.

The next chapter discusses arguments for free will.

CHAPTER 2

ARGUMENTS FOR FREE WILL

> Men at sometime, are Masters of their Fates.
> The fault (deer Brutus), is not in our Starres,
> But in our Selves . . . .
>
> Shakespeare, *Julius Caesar*[1]

This chapter surveys the arguments of representative philosophers, scientists, and others who have endorsed the concept of free will. It begins with the ancient Greek philosopher Aristotle, whose approach continues to be relevant more than two millennia after his lifetime. The section "Historical Dualism (or Nondualism)" describes the views of René Descartes, Immanuel Kant, and recent Christianity. As explained therein, one can cogently argue that Descartes and Kant were not, in the last analysis, dualists; nevertheless, dualism is the position for which they have long been famous. The section "Contemporary Philosophical and Scientific Arguments for Free Will" updates Aristotle and informs my own views on free will, which are elaborated in Chapter 3.

The Introduction to this book defined "free will" as "the independent ability to make conscious decisions that are neither predetermined nor random." Unless otherwise noted, the authors discussed below generally share this definition.

ARISTOTLE (384–322 BCE)

Aristotle did not use the term "free will." He did, however, discuss principles that are relevant to our contemporary discussions of free will. The most accessible treatments of those concepts can be found in his ethical writings.

In his *Nicomachean Ethics*, Aristotle wrote that "choice appears to be concerned with things that are up to us" (1111b30).[2] Choice, for Aristotle, was a subset (species) of a larger class (genus) of the voluntary. He discussed at some length the difference between the voluntary and the involuntary (1109b30–1111b3). What is involuntary is not, of course, "up to us," though it is not always easy to distinguish the involuntary from the voluntary. Choice "appears to be something voluntary" that is "accompanied by reason and thought" (1112a14–16). "Since what is chosen is a certain longing, marked by deliberation, for something that is up to us, choice would in fact be a deliberative longing for things that are up to us" (1113a10–13).

According to Aristotle, "choice" is a deliberative activity that excludes the relatively nonreflective actions of children and nonhuman animals: "both children and animals share in what is voluntary but not in choice, and we say that sudden actions are voluntary but do not stem from choice" (1111b9–11).[3] This differs somewhat from my own understanding. As explained in the Introduction, I regard even rapid decisions, with some qualifications, as choices resulting from free will. But this disagreement with Aristotle appears to be merely semantic: Aristotle defines choice as a deliberative, purposive activity, whereas I would define it to include quick choices and decisions.

Aristotle evidently did not anticipate the view (see Chapter 1) that all choices and decisions, whether deliberative or quick, are predetermined from the time of the Big Bang to the present (Aristotle, of course, had no inkling about the Big Bang, and he considered the universe to be eternal). However, by stating that even the sudden actions of children and nonhumans are "voluntary," Aristotle appeared to hold that their associated mental states are not predetermined. There is no indication that Aristotle's concept of the "voluntary" was the same as the latter-day compatibilist view that an action can be "voluntary" even though it is predetermined. After all, one of Aristotle's great contributions to philosophy was (following Plato's lead) the articulation of the principle of (non)contradiction.[4]

Aristotle wrote that humans do not deliberate about "things that are in motion but that always come into being in the same ways, whether from necessity or also by nature (or on account of some other cause)—for example, about solstices and sunrises. Nor does anyone deliberate about things that are different at different times—for example, droughts and rains—or about what arises from chance—for example, the discovery of treasure" (1112a24–28).

"But we do deliberate about things that are up to us and subject to action, and these are in fact what remain. For nature, necessity, and chance seem to be causes, but so too are intellect and all that comes about through a human being. When it comes to human beings, each deliberates about actions that come about through his own doing" (1112a31–39). "[D]eliberating occurs in matters that are for the most part so, where it is unclear how they will turn out and in which something is undetermined" (1112b8–9).

"It seems, then, as has been said, that a human being is an origin of his actions. Deliberation is concerned with actions that happen through one's own doing, and the actions are for the sake of something else" (1112b33).[5]

Aristotle is here discussing action, but he connects action in this context with choice and deliberation. He takes seriously the view that our conscious cogitation is something real that may result, in many cases, in actions. He distinguishes matters that are "up to us" from physical "nature, necessity, and chance," thereby precluding any notion that our thoughts, choices, decisions, and actions are predetermined or random.

For Aristotle, virtue and vice are "up to us" and, with some exceptions, such as being forced or acting in ignorance (when one is not responsible for his/her own ignorance), voluntary. "For in the cases in which it is up to us to act, so too is not acting; and where there may be a 'no,' there may also be a 'yes.' " Humans are the "origin and begetter" of their actions, just as they are the "origin and begetter" of their offspring. "But if these points appear to be the case, and we are not able to trace the origins [of our actions] to any other origins apart from those within us, then these very actions are up to us and voluntary." Accordingly, lawgivers punish even ignorant offenders when the ignorance is self-caused, e.g., in cases of drunkenness or carelessness (1113b6–34).[6]

This leads Aristotle to an important principle. The unethical person is like a sick person:

> At one time, then, it was possible for [a sick person] not to be sick; but in letting himself go, it is no longer possible, just as it is not possible for someone to toss away a stone and then retrieve it. Nonetheless, the throwing was up to him, because the origin [of the throwing] is in him. **In this way too, it was possible at the beginning for both the unjust person and the licentious one not to become such as they are, and hence they are what they are voluntarily; but once they become such, it is no longer possible for them to be otherwise.** (1114a17–21 [emphasis added, brackets in the Bartlett-Collins translation])[7]

I myself doubt that it is impossible for all such persons to recover an ethical perspective (see Chapter 3), but Aristotle's pessimism may be appropriate in some, perhaps many, cases.

Aristotle held that children and young adults learn ethical behavior through their upbringing and education. Their parents, teachers, and lawgivers try to instill good ethical habits in them. When they are older, this habituation makes it easier for them to adopt a reasoned approach to ethics. Unfortunately, some young people are habituated in the opposite manner and thus more easily succumb to vice. This makes it difficult for them to overcome their negative habituation and adopt an ethical attitude and behavior. For good or for ill, our habits lead to the development of our character or what Aristotle called "characteristics" (see 1103a1–1106a24). Summarizing Aristotle, translators and editors Robert C. Bartlett and Susan D. Collins have aptly defined what he meant by "characteristics" as follow: "Our characteristics, in this sense, display our character, the habits of body and mind that have been formed through habituation and that constitute a certain way of holding oneself toward the world, so to speak."[8]

Accordingly, Aristotle acknowledged the effects of good habituation leading to good character and bad habituation leading to bad character. He expressed opposition to the view that some people are just good or bad by nature—by their genes, as we would say today.[9] He concluded that virtue and vice are both voluntary, because "we ourselves are somehow joint causes of our characteristics, and by being of a certain sort, we set down this or that sort of end . . . ." (1114a32–1114b25).

Aristotle concludes his discussion of what we would call free will with the following summary:

> As for what concerns the virtues taken together, then, their genus was stated by us in outline: that they are means and that they are characteristics; that they are in themselves productive of the actions out of which they come to be; that they are up to us and voluntary; and that [they prompt us to act] in the way correct reason commands. But actions and characteristics are not voluntary in a similar way. For in the case of actions, from the beginning up to the end we exercise authoritative control over them, knowing the particulars involved; whereas in the case of the characteristics, we are in

control of the beginning of them, but at each moment, the growth [that results from the relevant activity] is not noticed, just as in the case of illnesses. But because it was once in our power to make use of [the characteristics] in this or that way, they are voluntary. (1114b26–1115a6 [brackets in the Bartlett-Collins translation])

In our own time, professional philosopher Robert Kane, an advocate of free will, has developed Aristotle's concept of "characteristics" into what he calls "self-forming actions" or "self-forming willings."[10] Kane formulates a theory of free will that involves both alternative possibilities and ultimate responsibility ("up to us"). Among other things, he calls our attention to the following passage in Aristotle's *Physics*: "The stick moves the stone and is moved by the hand, which is again moved by the man; in the man, however, we have reached a mover that is not so in virtue of being moved by something else."[11]

Like Kane, I think Aristotle's approach remains relevant in our era, as it addresses human nature that appears to be rather unchanging over the millennia. For discussions of how an individual human could be a mover without being moved by something else, see the section "Contemporary Philosophical and Scientific Arguments for Free Will" later in this chapter. See also Chapter 3, which, among other things, further addresses relevant aspects of Aristotle's philosophy.

HISTORICAL DUALISM (OR NONDUALISM)

René Descartes (1596–1650)

The writings of Descartes present a difficult interpretive problem. In his twenties, Descartes jotted the following in a private notebook: "Actors, taught not to let any embarrassment show on their faces, put on a mask. I will do the same. So far, I have been a spectator in this theatre which is the world, but I am now about to mount the stage, and I come forward masked."[12] He himself acknowledged suppressing some of his work after he became aware of the persecution of Galileo by the Inquisition in 1633.[13] Arthur M. Melzer comments: "When Descartes, . . . deeply shaken by the recent condemnation, imprisonment, and recantation of Galileo, expressed his desperate longing (in a letter to Mersenne) 'to live in peace and to continue the life [of philosophy that] I have begun,' he added that he would do so by following Ovid's ancient dictum: *bene vixit, bene qui latuit* (he has lived well who has remained well hidden)."[14] This was also the epitaph that Descartes chose for his tombstone.[15]

Thus, from the beginning to the end of his adult life, Descartes hinted at a need for concealment. This is not surprising considering the times in which he lived. One scholar has remarked on "the notorious caution of [Descartes's] manner of writing . . . ."[16] Another has noted several indications by Descartes himself as well as other philosophers that Descartes wrote in a manner intended to placate the reigning religious and political authorities while intimating his real views to philosophically prepared minds.[17]

Descartes is most famous today for his doctrine of mind-body dualism. He began his philosophical inquiry by doubting everything. He eventually concluded, however, that he could not doubt that "I think, therefore I am." This led him, by somewhat circuitous logic, to a fundamental duality between mind/soul and body.[18] "And accordingly, it is certain that I—that is, my soul, by which I am what I am—am really distinct from my body, and can exist without it."[19] Needless to say, this is orthodox Christian doctrine. Among other things, it is consistent with the Christian view that the soul survives death. In fact, Descartes stated that "while the body can very easily perish, the mind—or the soul of man, for I make no distinction between them—is immortal by its very nature."[20] It is an interesting question whether Descartes really believed this.[21] However, it is precisely this dualism between mind/soul and body that is associated with Descartes today—a medieval notion that scholars love to debunk.[22]

Many decades ago, Leo Strauss wrote: "We can barely allude to the question of Descartes's technique of writing, to a question which seems to baffle all his students because of the extreme caution with which that philosopher constantly acted."[23] Strauss also remarked that "if an author contradicts himself, the reader does well to suspend his judgment on what the author thought about the subject in question, and to use his powers rather for finding out by himself which of the two contradictory assertions is true."[24]

We find an example of Descartes's caution and self-contradiction in what he wrote on the subject of free will.

In section 37 of part 1 of his *Principles of Philosophy*, Descartes strongly supported free will:

> 37. *The supreme perfection of man is that he acts freely or voluntarily, and it is this which makes him deserve praise or blame.*
>
> The extremely broad scope of the will is part of its very nature. And it is a supreme perfection in man that he acts voluntarily, that is, freely; this makes him in a special way the author of his actions and deserving of praise for what he does. We do not praise automatons for accurately producing all the movements they were designed to perform, because the production of these movements occurs necessarily. It is the designer who is praised for constructing such carefully-made devices; for in constructing them he acted not out of necessity but freely. By the same principle, when we embrace the truth, our doing so voluntarily is much more to our credit than would be the case if we could not do otherwise.[25]

In section 38, Descartes argues that "[t]he fact that we fall into error is a defect in the way we act or in the use we make of our freedom, but not a defect in our nature."[26]

In section 39, he writes: "That there is freedom in our will, and that we have power in many cases to give or withhold our assent at will, is so evident that it must be counted among the first and most common notions that are innate in us."[27]

But then, after the reader concludes that Descartes strongly supports free will, our author does a sudden about-face:

> 40. *It is also certain that everything was preordained by God.*
>
> But now that we have come to know God, we perceive in him a power so immeasurable that we regard it as impious to suppose that we could ever do anything which was not already preordained by him. And we can easily get ourselves into great difficulties if we attempt to reconcile this divine preordination with the freedom of our will, or attempt to grasp both these things at once.[28]

In the very next section, Descartes purports to reconcile the contradiction:

> 41. *How to reconcile the freedom of our will with divine preordination.*
>
> But we shall get out of these difficulties if we remember that our mind is finite, while the power of God is infinite—the power by which he not only knew from eternity whatever is or can be, but also willed it and preordained it. We may attain sufficient knowledge of this power to perceive clearly and distinctly that God possesses it; but we cannot get a sufficient grasp of it to see how it leaves the free actions of men undetermined. Nonetheless, we have such close awareness of the freedom and

> indifference which is in us, that there is nothing we can grasp more evidently or more perfectly. And it would be absurd, simply because we do not grasp one thing, which we know must by its very nature be beyond our comprehension, to doubt something else of which we have an intimate grasp and which we experience within ourselves.[29]

There Descartes leaves us. It is submitted that his attempted reconciliation of the contradiction does not actually resolve it and that Descartes was perfectly aware of this failure. A similar ineffective attempt to explain the contradiction away is set forth in sections 146 and 147 of part 2 of his writing *The Passions of the Soul*.[30] Although Descartes also supports free will in his *Meditations on First Philosophy*,[31] his above-noted contradiction between free will and divine preordination seems to leave the reader who asserts free will with no option but to reject the historical legacy of divine predeterminism that had been bequeathed to subsequent ages by the later writings of Augustine and resurrected and amplified by Luther and Calvin in the sixteenth century.[32]

Indeed, Thomas Hobbes, in his Twelfth Objection to Descartes's *Meditations on First Philosophy*, argued that "the freedom of the will is assumed without proof, and in opposition to the view of the Calvinists."[33] Descartes responded: "Our freedom is very evident by the natural light. . . . There may indeed be many people who, when they consider the fact that God pre-ordains all things, cannot grasp how this is consistent with our freedom. But if we simply consider ourselves, we will all realize in the light of our own experience that voluntariness and freedom are one and the same thing. This is no place for examining the opinion of other people on this subject."[34]

This appears to be Descartes's final judgment on free will: we must trust our own "natural light" (correct reason, unassisted by revelation[35]) and "experience" over received theological dogmas, including but not limited to the dogmas of such "other people" as Calvin and his followers. As for Hobbes, see the discussions of his views in the preceding chapter.

Immanuel Kant (1724–1804)

Immanuel Kant held that "[o]ur cognitive faculty as a whole has two domains: that of the concepts of nature [which he called the 'phenomenal' realm] and that of the concept of freedom [which he called the 'noumenal' realm]." The phenomenal sphere is the locus of perception and, ultimately, what we now call "scientific" knowledge; it involves only "appearances," for which we cannot know the ultimate reality. The noumenal dimension concerns free will, morality, the soul, and God—the "things-in-themselves," according to Kant.[36]

By Kant's time, there was an intellectual contest between those classical, Newtonian physicists (and their philosophical allies) who saw everything as governed by intractable and predetermined laws of nature and the philosophers and theologians who thought that human beings had free will, a soul, and/or an afterlife. Today, that scholarly debate has devolved into a culture war between scientists and religionists with all kinds of political implications and consequences. Kant had hoped to prevent such violent conflict by postulating that the empirical phenomena of natural science are different from such noumenal facts as free will and morality. In the twentieth century, special and general relativity and quantum physics challenged some of the fundamental premises of classical physics. In recent decades, as discussed elsewhere in this and the next chapter, neuroscientists such as Peter Ulric Tse and William R. (W. R.) Klemm, quantum physicists such as Henry P. Stapp, and chaos and complexity theorists such as physicist Ilya Prigogine and biologist Stuart A. Kauffmann have argued that correct scientific principles and evidence actually support free will rather than rebut it. One wonders how Kant would have reacted to such developments.[37]

Kant postulated that what he called the "categorical imperative" is the fundamental law of all morality: "There is, therefore, only a single categorical imperative and it is this: *act only in accordance with that maxim through which you can at the same time will that it become a universal law.*"[38] "For the purpose of achieving this it is of the utmost importance to take warning that we must not let ourselves think of wanting to derive the reality of this principle from the *special property of human nature*. For, duty is to be practical unconditional necessity of action and it must therefore hold for all rational beings (to which alone an imperative can apply at all) and *only because of this* be also a law for all human wills."[39] Nonhuman "rational beings," for Kant, included rational beings on other planets, God, and, possibly, angels.[40]

For Kant, the metaphysically decisive proof of free will was its integrality to the moral law (the categorical imperative).[41] However, "we do not indeed comprehend the practical unconditional necessity of the moral imperative, but we nevertheless comprehend its *incomprehensibility*; and this is all that can fairly be required of a philosophy that strives in its principles to the very boundary of human reason."[42] In his later *Metaphysics of Morals*, he concluded that "these laws [the categorical imperative and freedom of choice], like mathematical postulates, are *incapable of being proved* and yet *apodictic*...."[43] This appears to be Kant's final word on the subject.

Kant argued that we need to accept the paradox that our actions are determined, as a matter of appearance, by preceding events in the phenomenal world but that we, nevertheless, have free will transcending time in the noumenal realm.[44] Astrophysicist-philosopher Bob (Robert O.) Doyle has explained Kant's position as follows:

> Kant's *noumenal* world outside of space and time is a variation on Plato's concept of Soul, Descartes' mental world, and the Scholastic idea of a world in which all times are present to the eye of God. His idea of free will is a most esoteric form of compatibilism. Our decisions are made in our souls outside of time and only *appear* determined to our senses, which are governed by our built-in *a priori* forms of sensible perception, like space and time, and built-in categories or concepts of intelligible understanding.[45]

This "timeless agency" interpretation is not, however, the only understanding of Kant on offer.

Robert Hanna and Henry E. Allison, for example, have discussed other interpretations of Kant.[46] Hanna argues—mostly, but not entirely, on the basis of Kant's postcritical works—that Kant developed a nondualistic, biological theory of freedom.[47]

Kant rejected divine (pre)determinism.[48] But his philosophy of history taught that providence (nature) drives human progress largely through passions, not virtues. He was familiar with Adam Smith's *Wealth of Nations*,[49] which contained similar views. Among other things, Kant advised that the frequent warfare and power struggles among nation states could ultimately lead, by some kind of providential logic, to a league of nations and perpetual peace. Accordingly, it might appear that Kant replaced the ancient emphasis on the importance of intellectual and moral virtue in political leadership (see, for example, Plato's *Republic* and *Seventh Letter*) with a modern view that the blind conflict of competing passions ultimately has beneficent effects. Virtue is good on an individual level, but it is not of decisive importance in the overall political scheme of things. Predeterminism somehow supersedes individual free will on the grand historical stage.[50] Nevertheless, in one of his latest works, Kant taught that perpetual peace should be sought, regardless of its attainability, as a moral duty pursuant to metaphysical principles.[51] Thus, Kant did not entirely give up on the importance and efficacy of moral effort in the political sphere.

Recent Christianity

As explained in Chapter 1, Paul, Augustine, Luther, Calvin, and Hobbes all taught one or more forms of theological predeterminism or predestinarianism. According to one author, Luther later rejected, at least in part, his early views favoring predestination and denying free will, following the changing viewpoint of his close associate, Philipp Melanchthon.[52] However, Luther never stated that his views on predestination had changed, and there is evidence that they did not change.[53] Similarly, Professor Alister E. McGrath sees the change in Lutheran predestinarian doctrine as occurring more with Melanchthon than with Luther; McGrath also notes the controversy about this issue in Lutheran circles after Luther's death.[54] McGrath adds that "later Lutheranism marginalized Luther's 1525 insights into divine predestination, preferring to work within the framework of a free human response to God, rather than a sovereign divine election of specific individuals. For later sixteenth-century Lutheranism, 'election' meant a human decision to love God, not God's decision to elect certain individuals."[55] Nevertheless, debates over predestination have continued in recent times among American denominations of Lutheranism, with at least one denomination insisting on a version of Luther's original predestinarian teaching.[56]

The developments in Lutheranism in the direction of free will have been similar to developments in Christianity generally.[57] Under this new conception, salvation is decided not by God but by each person's free will, residing in an individual, nonphysical soul. In one version, salvation depends on the person's faith in doctrinal matters (the traditional Protestant doctrine of salvation by faith, not works, but with faith being up to the individual and not predetermined by God as a matter of grace). In another, a person's eternal fate rests on moral conduct during life: those who are sufficiently moral will go to heaven, the rest to hell. The tying of salvation to morality is the opposite extreme of the predestinarian principles of Luther and Calvin. Yet it is a quite popular notion, reminiscent of the Myth of Er in Book 10 of Plato's *Republic*. It is ironic that the pagan Plato was ultimately more successful on this issue than the most famous Protestant theologians.

CONTEMPORARY PHILOSOPHICAL AND SCIENTIFIC ARGUMENTS FOR FREE WILL

Strong arguments supporting free will emerged during the early decades of the twenty-first century. These arguments eschewed traditional dualism and attempted to establish a secular philosophical and/or scientific basis for free will. The present section explores some of these interesting accounts.

Robert O. (Bob) Doyle

Bob Doyle[58] is an astrophysicist and a philosopher who supports the concept of free will.[59] Among many other things, he makes the following arguments/observations.

Free will is the product of evolution by natural selection, starting with such lower forms of life as bacteria (which have what he calls "behavioral free will") and ending with a more developed free will in humans.[60] For heuristic purposes, he teaches a two-stage temporal model of free will: the first stage is an indeterministic random generator of alternative possibilities (freedom), and the second stage is the deliberate choice by the agent from among the possibilities (will).[61] In reality, however, "the random generation stage is going on constantly, driven by internal proprioceptions and external environmental perceptions."[62] The interaction between the indeterministic stage and what he calls the stage of "adequate determination" or "adequate determinism" enables free will.

Although Doyle uses the term "adequate determinism" to describe his second stage, he rejects predeterminism, which he defines as "the idea that a strict causal determinism is true, with a causal chain of events back to the origin of the universe, and one possible future."[63] The indeterministic random generation stage, which is rooted (at least in part) in quantum mechanics, interrupts any presumed inevitable sequence of cause and effect and thereby invalidates any theory of predeterminism.[64]

Doyle's two-stage model avoids the false dichotomy discussed in the Introduction. That "standard argument" against free will posits that the choice is between absolute (pre)determinism and absolute indeterminism, neither being consistent with free will. Instead, Doyle follows the two-stage model of philosopher-psychologist William James (1842–1910): "James accomplishes this by using chance simply to create genuinely new and unpredictable alternatives for action, following which a choice can be made by a will that is consistent with character, values, and especially with one's desires and feelings, which James considered an essential part of the will."[65] The following subsections discuss the neuroscience supporting this paradigm.

William R. (W. R.) Klemm

Neuroscientist William R. Klemm's book *Making a Scientific Case for Conscious Agency and Free Will* and his two earlier books bearing on the scientific basis of consciousness and free will[66] are the culmination of decades of research and writing. In addition to a large number of books, he has authored over fifty book chapters in edited collections and hundreds of peer-reviewed papers on these and related subjects.

Klemm begins *Making a Scientific Case for Conscious Agency and Free Will* with an operational definition of free will: "Free will occurs when a person makes a conscious choice from multiple options, none of which are predetermined or compelled."[67]

He asks: "If human consciousness is the product of evolution, why would it not have not have agency? Natural selection could be expected to favor evolution of both conscious agency and free will. Both are useful in the complex decisions that humans are capable of making . . . ."[68] Section 3.4 of his book discusses these thoughts in greater depth.[69]

Klemm elaborates at length his view that the brain itself—which consists of about 86 billion neurons, each of which can have hundreds of connections—gives rise, through quite complicated processes, to consciousness and free will. He discusses both his theory and those of other neuroscientists with regard to how such emergence could occur.

Klemm does not claim infallibility for his or any other theory. Rather, he observes that much more research needs to be done. He concludes the book with the following remarks: "Hasn't the time arrived when we should generate the hypothesis that humans do have a meaningful degree of free will and devise controlled experiments to test that hypothesis? We will have to look to neuroscience for enlightenment on this subject—no, not the neuroscience of the 1980s [Libet et al.] but the neuroscience of neural networks that is yet to come."[70]

Professor Klemm has published widely on many matters. I express no opinion regarding his views about subjects other than the interrelationship between neuroscience and free will.

Peter Ulric Tse

The following discussion addresses neuroscientist Peter Ulric Tse's 2013 book *The Neural Basis of Free Will: Criterial Causation* and his 2018 book chapter titled "Two Types of Libertarian Free Will Are Realized in the Human Brain."[71] Parenthetical textual references to *Neural Basis* are to section/subsection number (each paragraph of the body of the book is identified with a subsection designation), and references to "Two Types of Libertarian Free Will" are to page numbers.

Tse begins but does not end with a commonsense observation:

> The physical realizations of consciousness and mental causation have already been "solved" by evolution. If a learned man asserts that you cannot sail westward from Portugal to reach China because the world is flat, and an explorer accomplishes this nonetheless, which one needs to change his view? The solution to the mind-body problem will not come from logical

derivations that yield conclusions that follow from unproven assumptions. What if an assumption of reductionism is wrong? What if an assumption of determinism is wrong? The solution to the ancient mind-body problem will eventually emerge as scientists unravel the way that mind is in fact realized in neuronal activity at the level of interneuronal and intraneuronal circuits. New options concerning free will, mental causation, and the mind-body problem can and will emerge in light of neuroscientific data. (*Neural Basis*, § 1.17)

Neural Basis contains many references to the evolutionary development of free will (§§ 3.8. 4.5, 6.22, 7.11, 10.7–10.11, 10.45–10.47). Similarly, in "Two Types of Libertarian Free Will," Tse comments: "We have no choice but to have a libertarian free will because evolution fashioned our nervous systems to have it. Those animals that had a nervous system that realized a libertarian free will survived to the point of procreation better than those that did not" (179; see also 181, 184, 189). (Note: Tse's use of the word "libertarian" does not refer to a political movement but rather to a philosophical position supporting free will.)

Tse does not accept the notion that there is a separate, immaterial "soul" in the human physical body—the (in)famous "ghost in the machine" (*Neural Basis*, § 7.15). He thinks that the question of free will stands or falls on whether free will evolved by way of Darwinian natural selection and emergence; his procedure is strictly scientific, not religious (*Neural Basis* passim).

But Tse rejects some views commonly associated with physicalism. For example, the doctrine of eliminative materialism or epiphenomenalism holds that everything proceeds by way of causal determinism resulting from the motions and interactions of the most elementary particles as revealed by the latest investigations or speculations of particle physicists. Everything beyond such particle interactions is allegedly epiphenomenal, that is, noncausal. In other words, all activity of all inorganic and organic things is predetermined by a succession of particle-level causes and effects going back to the beginning of the universe. Every human event—including every human social, economic, political, and intellectual event—is, by this logic, predetermined, and our notions of free will are purely illusory. Tse shows that such a view is unscientific in that it ignores many aspects of nature. Among other things, it does not account for the critical importance of information, which has neither matter nor energy and accordingly is not subject to the first and second laws of thermodynamics. Tse's concept of criterial causation essentially involves information, but the eliminative materialists dismiss that fact just as they claim that the informational aspects of genetics are epiphenomenal (*Neural Basis*: see index references for "Epiphenomenalism" and "Information").

Furthermore, Tse observes that most scientists today accept the findings of quantum physics—including ontological (actual) indeterminism, as distinguished from particle determinism. He argues that the brain, through rather complex chemical (neurotransmitter) and other mechanisms, harnesses indeterminism to create the options necessary for free will. He summarizes his essential argument in section 7.23 of *Neural Basis*:

> I have proposed a three-stage model of a neuronal mechanism that underlies mental causation and free will, according to which (1) new physical/informational criteria are set in a neuronal circuit on the basis of preceding physical/mental processing at t1, in part via a mechanism of rapid synaptic resetting that effectively changes the inputs to a postsynaptic neuron. These changes can be driven volitionally or nonvolitionally, depending on the neural circuitry involved. (2) At t2, inherently variable inputs arrive at the postsynaptic neuron, and (3) at t3 physical/informational criteria are met or not met, leading to postsynaptic neural firing or not. Such "criterial causation" (§6.8) is important because it allows neurons to alter the physical realization of future mental events in a way that escapes the problem of mental-on-physical self-causation, a problem that has been at the root of basic criticisms of the possibility of mental causation (§§A2.2–A2.5) and free will (§§7.1–7.5). Substantial theoretical work, however, is still required to provide a more complete account of free will and mental causation. One could argue that neuronal criterial causation and ontological indeterminism are at best necessary but cannot alone be sufficient for the physical realization of mental causation and a strong

free will, because without a conscious agent with intentional states that could exploit the three-stage process to fulfill its ends, we might be no more than mindless, unpredictable "zombies"....

The final three chapters of *Neural Basis* discuss the concerns identified in the last sentence of the preceding quotation.

Tse's writing is often difficult and technical, but it is possible even for someone (like me) who is not a neuroscientist to follow the main thread of his argument. Tse is never dogmatic. He alerts the reader to the fact that more work is necessary, and he identifies the types of further inquiries that are appropriate. He eloquently summarizes some of the important remaining issues in the final paragraph (§ 0.4) of the preface to his book:

The deepest problems have yet to be solved. We do not understand the neural code. We do not understand how mental events can be causal. We do not understand how consciousness can be realized in physical neuronal activity. We have yet to be liberated by a Darwin who will provide an overarching conceptual framework that unifies all the data in the field. We have yet to find our Watson and Crick who will crack the neural code. It is truly a wonderful time to be in the brain sciences because we know so much, and yet the deepest issues remain wide open. We find ourselves in the midst of one of humankind's greatest ages of exploration and discovery—only it is not exploration of the outer world with ships and sextants. It is exploration of the inner world with scanners and electrodes.

One wonders what will be understood about the human brain a hundred years from now. One thing is clear: Tse has made a substantial contribution to knowledge in this important field. The present account has only scratched the surface of his in-depth analysis.

Christian List

Christian List is a professor of philosophy and political science and a Fellow of the British Academy. In his book *Why Free Will is Real*,[72] he defends the concept of free will against three attacks: (1) radical materialism, which denies that humans are intentional agents; (2) (pre)determinism, which denies that humans have alternative possibilities; and (3) epiphenomenalism, which denies that humans have causal control over their actions. Although his entire argument is complicated, List's main claim is that the opponents of free will commit a category error in attempting to reduce all human thought and behavior to (classical) physics. He argues that much of human thought and action cannot be explained at the physical level. The "special sciences" (the studies of biomedical, social, economic, and political phenomena, for example) exhibit the existence and operation of higher-level human thought and activity, which is impossible to understand on the basis of physics alone. He suggests that higher-order human activities involving free will **emerged** from a lower-level physical substrate, though he does not attempt to explain how such emergence and/or evolution occurred. For relevant theories of emergence/evolution, see, for example, the writings of neuroscientists William R. (W. R.) Klemm, and Peter Ulric Tse, discussed above, and neuroanthropologist Terrence W. Deacon's book *Incomplete Nature: How Mind Emerged from Matter*.[73]

Henry P. Stapp

Quantum physicist Henry P. Stapp worked with two of the most famous twentieth-century quantum physicists—Werner Heisenberg and Wolfgang Pauli. After many decades of study, experience, and writing in this field, Stapp published a book titled *Quantum Theory and Free Will: How Mental Intentions Translate into Bodily Actions*.[74] He has also published other books and articles relevant to quantum physics and free will.[75]

Stapp sets forth his foundational argument on page 5 of *Quantum Theory and Free Will*:

> The thesis expounded in this book is that von Neumann's orthodox formulation of quantum mechanics, elucidated where needed by the ideas of Heisenberg, Dirac, Wheeler, and the mathematician, logician, and philosopher Alfred North Whitehead, and updated to the relativistic form developed by Tomonaga and Schwinger, can be regarded as a theory of reality that is sufficiently detailed and accurate to deal with the issues of the general nature of our mental aspects, and of the causal connection of our conscious minds to the material world

in which our brains and bodies are embedded.

Stapp distinguishes between the classical physics of Newton, Laplace, and many other scientists and philosophers during the last few centuries, on the one hand, and the quantum physics discovered and elaborated in the twentieth century, on the other. The conventional wisdom is that quantum mechanics is confined to microscopic phenomena and that "laws" of classical physics govern all macroscopic events, including the physical operations of the brain. From this premise, many scientists and philosophers even today deny that quantum mechanics has anything to do with the brain. Rather, many assert, the brain operates by materialistic cause and effect in the manner of classical physics. Thus, the argument goes, every brain event is predetermined by physical causation from the beginning of time. Consciousness and free will, in this view, are merely illusory and "epiphenomenal" (noncausal).

In contrast, Stapp argues that "[t]he quantum resuscitation of the causal power of our thoughts overturns the absurd classical notion that nature has endowed us with conscious minds whose only power and function is to delude us into believing that it is helping us to create a future that advances our felt values, while in actuality that future was predetermined 15 billion years ago" (26–27).

Stapp totally rejects the common view that quantum mechanics is limited to microscopic events. This incorrect premise has been the source of a widespread pernicious belief that quantum mechanics has little or nothing to do with the big questions of the basic nature of the world, and of ourselves. That belief inspires the related notion that the consideration of quantum effects can be relegated to specialists who are interested in the atomic minutiae, while thinkers concerned with the big human issues can pursue their thinking (apart from the intrusion of the quantum elements of random chance) within the simpler framework of classical physics, which excludes our minds from the causal dynamics. But, according to quantum mechanics, the inclusion of the effects of our mental intentions upon the <u>macroscopic</u> behavior of our brains and bodies is absolutely essential to a correct understanding of the dynamical role of our human minds in workings of nature, and hence to a valid self image. (73–74 [underlining in the original]; see also 13, 35–36, 59–60, 65–66)

Stapp's basic idea is that the brain is constantly projecting—pursuant to quantum principles—alternative possibilities. The "ego" (self, observer) chooses one of these possibilities, and nature then responds either affirmatively or negatively. If nature's answer is affirmative, that possibility is actualized by a collapse of the wave function and the elimination from the brain's history of the roads not taken. This understanding is, of course, incompatible with the notion of classical mechanics and predeterminism that no alternative possibilities exist. Stapp's argument for free will rests on the premise that alternative possibilities are, on the basis of quantum theory and the biological mechanism of ion channels in the brain (13), an ontological reality.

I do not have sufficient expertise in physics and mathematics to evaluate the technical details of Stapp's account of quantum mechanics, but his general argument in support of free will and his deconstruction of scientific predeterminism are revolutionary and significant. I must conclude with the caveat that the ultimate proof of his theory depends, according to Stapp himself, on mathematical proofs that are beyond my ken.

Jeffrey M. Schwartz

Jeffrey M. Schwartz, MD, a research professor of psychiatry, is the principal author of *The Mind and the Brain: Neuroplasticity and the Power of Mental Force*[76] and other books and writings.

Schwartz has been a longtime intellectual friend and collaborator of Henry Stapp. He has supplemented Stapp's work on quantum mechanics by, among other things, applying it in the context of neuroplasticity. Neuroplasticity, in Schwartz's words, "refers to the ability of neurons to forge new connections, to blaze new paths through the cortex, even to assume new roles. In shorthand, neuroplasticity means rewiring of the brain."[77]

Schwartz was the lead author of an academic paper titled "Quantum Physics in Neuroscience and

Psychology: A Neurophysical Model of Mind-Brain Interaction." Henry Stapp and neuroscientist Mario Beauregard were coauthors.[78] The article explains that the term "self-directed neuroplasticity" is "a general description of the principle that focused training and effort can systematically alter cerebral function in a predictable and potentially therapeutic manner" (1310).

This approach is based on what the article calls "contemporary" (quantum) physics rather than pre-quantum classical physics. Under "the classic conception of the world, all causal connections between observables are explainable in terms of mechanical interactions between material realities" (1310). Classical physics cannot explain or account for "feeling," "knowing," and "effort" (1310). In contrast, quantum physics can, according to this article, provide a framework for understanding such phenomena.

Unlike classical physics, quantum physics rejects the notion that all aspects of experience are consequences solely of physical brain mechanisms. Classical physics treats the mind as a passive observer; it does not understand and cannot account for the mind as an active agent. Indeed, classical physics teaches that all mental events are mechanistically determined by micro-local physical brain activity. This reductionist approach is disproven by quantum physics, which demonstrates that quantum indeterminacy makes such determinism impossible. "The core phenomena necessary for the scientific description of self-directed neuroplasticity are processes that cannot be elaborated solely in terms of classic models of physics" (1311–13, 1324).

Neuroimaging studies show that the brain's prefrontal cortex is activated when the subject is engaged in willful mental activity, self-initiated activity, focused dispassionate self-observation, and self-directed regulation of emotional response. It takes "effort" ("willful training and directed mental effort") to redirect "the brain's resources away from lower level limbic responses and toward higher level prefrontal functions—and this does not happen passively" (1311–13).

Classical physics provides two alternative explanations for our first-person consciousness (our thoughts, ideas, and feelings). The first view is that such mental events are merely epiphenomenal by-products of the activity of our physical brains. Our "mental side is then a causally impotent sideshow that *is produced, or caused*, by your brain, but that *produces* no reciprocal action back upon your brain. The second way is to contend that each of your conscious experiences—each of your thoughts, ideas, or feelings—is the *very same thing* as some pattern in your brain" (1313–14 [italics in the original]). The authors of this paper reject both of these alternatives for detailed reasons that are set forth in the essay.

This paper goes into considerable depth in explaining how quantum physics interacts with neuroscientific phenomena in order to create indeterministic alternative templates for action, from which the observer chooses (and, when appropriate, acts upon) in the exercise of his/her free will (1315–25). It concludes by examining some other quantum interpretations (1325–26).

Jeffrey Schwartz discussed his neuroplasticity research and practice in his book *The Mind and the Brain: Neuroplasticity and the Power of Mental Force*.

Chapter 2 of this book explains how Schwartz began developing his approach in the late 1980s and 1990s with the diagnosis and treatment of patients who had obsessive-compulsive disorder (OCD). Neuroimaging showed that the OCD symptoms were caused by malfunctions in the brain's orbital frontal cortex and basal ganglia circuitry (74). A brain region called the striatum plays an important role in the development of habits (70). The habits connected with OCD could be changed by a four-step process, by which patients consciously learn to relabel, reattribute, refocus, and revalue OCD urges. After repeated habituation in this four-step process, the patients' neuroimaging results showed that their brains had significantly changed in a positive manner: "Done regularly, Refocusing strengthens a new automatic circuit and weakens the old, pathological one—training the brain, in effect, to replace old bad habits programmed into the caudate nucleus and basal ganglia with healthy new ones. When the focus of attention shifts, so do patterns of brain activity" (90–91). This neuroplastic brain change corresponds with the patients' being able to overcome their formerly strong OCD urges.[79]

These results led Schwartz to a conclusion relevant to the question of free will:

> The results achieved with OCD supported the notion that the conscious and willful mind differs from the brain and cannot be explained solely and completely by the

matter, by the material substance, of the brain. For the first time, hard science—for what could be "harder" than the metabolic activity measured by PET scans?—had weighed in on the side of mind-matter theories that ... question whether mind is nothing but matter. The changes the Four Steps can produce in the brain offered strong evidence that willful, mindful effort can alter brain function, and that such self-directed brain changes—neuroplasticity—are a genuine reality. (93–94)

As Schwartz observed, "This was the first study ever to show that cognitive-behavior therapy—or, indeed, any psychiatric treatment that did not rely on drugs—has the power to change faulty brain chemistry in a well-identified brain circuit. What's more, the therapy had been self-directed, something that was and to a great extent remains anathema to psychology and psychiatry" (90).

Schwartz concluded that "mental force," not reducible to the physical brain, arises from "willful effort": "What mental force does is activate a neuronal circuit. Once that new circuit begins to fire regularly, an OCD patient does not need as much effort to activate it subsequently; the basal ganglia, responsible for habitual behaviors, take care of that" (95). Consider how compatible this statement is with the analysis of Aristotle, discussed in the first section of the present chapter.

Before 1990, most neuroscientists believed that neuroplasticity only occurred in the brains of infants or young children. A number of studies have established, however, that neuroplasticity exists also in the adult brain. By the end of the 1990s, this fact was beginning to become generally accepted (253–54). *The Mind and the Brain* discusses many case studies that confirm, with scientific evidence, the existence of adult neuroplasticity—especially self-directed, volitional neuroplasticity.

However, mainstream neuroscience has refused to recognize that volition (free will) has anything to do with neuroplasticity. Schwartz explains:

> In neuroscience, . . . to take an interest in the role of volition and the mental effort behind it, and further to wonder whether volition plays a critical role in brain function, is virtually unheard of. Piles of brain imaging studies have shown that volitional processes are associated with increases in energy use in the frontal lobes: "right here," you can say while pointing to the bright spots on the PET scan, volition originates. But the research is mute on the chicken-and-egg question of what's causing what. Does activity in the frontal lobes cause volition, or does volition trigger activity in the frontal lobes? If the former, does the activity occur unbidden, as a mere mechanical resultant, or is it in any sense free? Generally, neuroscientists assume that the brain causes everything in the mind, period—further inquiry into causality is most unwelcome. (294)

Schwartz argues that "the paltry 35,000 or so genes in the human genome fall woefully short of the task of prescribing the wiring of our 100-trillion-synapse brain" (366). Elaborating on the technical (and less accessible) accounts of his quantum physicist friend Henry Stapp, Schwartz explains how quantum mechanics makes free will possible:

> In the brain, the flow of calcium ions within nerve terminals is subject to the Heisenberg Uncertainty Principle. There is a probability associated with whether the calcium ions will trigger the release of neurotransmitter from a terminal vesicle—a probability, that is, and not a certainty. There is, then, also a probability but not a certainty that this neuron will transmit the signal to the next one in the circuit, without which the signal dies without leading to an action. . . . [A] synapse exists as a superposition of the states "Release neurotransmitter" and "Don't release neurotransmitter." This superposition corresponds to a superposition of different possible courses of action. . . . By choosing whether and/or how to focus on the various possible states, the mind influences which one of them comes into being. (357)[80]

Schwartz concludes: "Through this mechanism, the mind can enter into the causal structure of the brain in a way that is not reducible to local mechanical processes . . . . This power of mind gives our thoughts efficacy, and our volition power" (360–61).[81]

Norman Doidge

Norman Doidge, MD, is a psychiatrist, psychoanalyst, and researcher. In his writings, Doidge acknowledges the contributions of Jeffrey Schwartz to the theory and practice of neuroplasticity. However, Doidge states that he does not have the expertise to evaluate the quantum physics theories of Schwartz (and, implicitly, of Stapp).[82]

Doidge's essay "Neuroplasticity, Perfectibility, and Three Ideas of Nature" (hereafter the "essay")[83] discusses the philosophical background of his neuroplasticity research and practice. He begins the essay by defining neuroplasticity as "the property of the brain that allows it to change its very structure and function in response to thought or mental experience" (443). He describes three phases of Western philosophical thinking about the brain. First, the ancient Greeks considered the universe to be analogous to an organism. "And, though the ancients often distinguished between mind and matter, they did not argue that different laws of nature applied to each, so there was not an explanatory gap as wide as would later emerge" (453). Second, modern philosophy introduced the concept of nature as one vast mechanism. Although Descartes probably disguised his philosophy in order to escape persecution from religious authorities, he became known for the so-called Cartesian dualism between mind and matter. The brain, in the modern view, was simply another type of machine. As Western civilization became more secular, this concept developed into mechanical determinism and the denial of free will. The third phase was inaugurated by Comte de Buffon (Georges Louis Leclerc) and Jean-Jacques

Rousseau. Rousseau famously argued that humans were "perfectible," by which he meant that the human brain was malleable. Education thus became of great importance for Rousseau, as evidenced by his work *Emile, or On Education* (1762). Doidge states that "I know of no one who predates Rousseau in making the claim that the brain's organization is plastic [malleable] in this way, and responsive to exercise." (458). In fact, Rousseau "was arguably the first thinker to begin reflecting on what plasticity meant for politics. . . . I would argue that neuroplasticity is the neurological correlate of perfectibility" (459).

Our contemporary brain-scanning technology demonstrates that, contrary to earlier conventional wisdom, certain brain functions are not strictly limited to particular areas of the brain. Many studies have shown that individuals who have brain damage in one area of their brains can, through training, cause another area of the brain to do the work formerly done by the damaged area. However, "[t]o argue that there is plasticity throughout the brain is not to argue that the brain is infinitely plastic. I do not wish to correct an overly rigid view of the brain with an overly flexible, neuro-utopian one. In fact, an infinitely plastic brain, like most utopias, would become a dystopian regime very quickly, for there must be a balance in the brain between its ability to change and its ability to hold on to what has proven to be worthwhile" (447).

"After Rousseau, the idea of perfectibility quickly got tied to the idea of 'progress.' " (460). The combined notion of limitless perfectibility and inevitable progress led to the horrors of the French and Russian revolutions. "Of course, the teaching of the neuroplasticitians is not that human beings are infinitely perfectible. As I have argued, the plastic paradox teaches that neuroplasticity is also responsible for rigid behaviors, and even some pathologies, along with all the potential flexibility that is within us" (461). "The same plasticity that gives rise to flexibility and our ability to modify our brains, also acts as a very conservative force preserving early tendencies. This means that political expectations that people can change radically, and quickly, are based on an inadequate understanding of how neuroplasticity works. But neither can a serious political philosophy count on human beings remaining rigidly predictable creatures, given our plasticity" (462).

Although Doidge does not thematically treat the question of free will, it is clear that he rejects (pre)determinism and accepts some version of free will (444, 455, 457). "Thus we are not, it turns out, merely the galley slaves of our selfish-gene masters, because thought gives us a significant degree of control over the expression of our genes, which then allows us to sculpt our central nervous system's microanatomy" (446). "[T]he neurons that process emotion, perception and cognition . . . are, indeed, neuroplastic, and subject to change, in the right circumstances. The driver of those changes turns out to be human thought itself" (462).

Doidge's book *The Brain That Changes Itself* and his later book titled *The Brain's Way of Healing: Remarkable Discoveries and Recoveries from the Frontiers of Neuroplasticity*[84] elaborate in great detail a wide variety of case studies showing how neuroplasticity actually works in clinical practice.

Summary

This chapter has considered the arguments of several proponents of free will, from Aristotle in ancient Greece to Descartes and Kant in modern times to some denominations of recent Christianity, and, finally, to some contemporary thinkers who argue that free will has a secular philosophical or scientific basis. There are many other interesting philosophical and scientific arguments for free will, but space does not permit a comprehensive treatment in this book. The following chapter presents my own views on this subject.

CHAPTER 3

FREE WILL AND HUMAN NATURE

> Two roads diverged in a yellow wood,
> And sorry I could not travel both
> And be one traveler, long I stood
> And looked down one as far as I could
> To where it bent in the undergrowth;
> Then took the other, as just as fair,
> And having perhaps the better claim,
> Because it was grassy and wanted wear;
> Though as for that the passing there
> Had worn them really about the same,
> And both that morning equally lay
> In leaves no step had trodden black.
> Oh, I kept the first for another day!
> Yet knowing how way leads on to way,
> I doubted if I should ever come back.
> I shall be telling this with a sigh
> Somewhere ages and ages hence:
> Two roads diverged in a wood, and I—
> I took the one less travelled by,
> And that has made all the difference.
>
> Robert Frost, "The Road Not Taken" (1916)

This famous poem metaphorically describes the choices we make in life. It underscores that every decision leads to consequences that, in turn, lead to other decisions with their own consequences. When, near the end of one's life, one thinks about the decisions one has made, one can see both how some good decisions led to good consequences and how some bad decisions led to bad consequences. More surprisingly, some decisions that appeared to be good at the time led, through lack of perfect knowledge, to bad consequences, and vice versa. This panorama of choices, decisions, and consequences is essential to human nature. The point about free will is that we do, or should do, the best we can in navigating the various circumstances that we face in life, making decisions that seem best to us at each moment.

But some claim that the person confronting the fork in the road is predetermined by God or Nature to take one of the paths and not the other. Chapter 1 of this book addressed that view. Chapter 2, in contrast, discussed perspectives of proponents of free will. The present chapter elaborates my own views on free will and human life. As noted in the Introduction, an argument can be made that nonhuman animals—or at least some of them—have varying degrees of free will. I actually agree with that proposition. This book focuses, however, on questions concerning the existence and significance of **human** free will.

THE CASE AGAINST THEOLOGICAL PREDETERMINISM OR MIRACULOUS INTERVENTION

The first section of Chapter 1 addressed theological predeterminism. At the end of that discussion I explained my reasons for rejecting divine predeterminism. For similar reasons I do not accept the view that a supreme being miraculously interferes with human thought or action on an ad hoc basis. There is no need for further elaboration of these matters.

THE CASE AGAINST SCIENTIFIC PREDETERMINISM, AD HOC DETERMINISM, AND COMPATIBILISM ("SOFT DETERMINISM")

Many opponents of free will dogmatically assert, as if they had ironclad scientific proof, that free will does not exist. Chapter 1 revealed the logical fallacies and scientific errors of typical representatives of scientific predeterminism, ad hoc determinism, and compatibilism/soft determinism. The section on ad hoc determinism in that chapter does not require further elaboration in the present section, though some of those issues will be revisited in the final section of this chapter. The immediately following discussion contains some additional remarks about scientific predeterminism and compatibilism/soft determinism.

A fundamental problem with scientific predeterminism is that science itself has disproved its validity. Scientific predeterminism is rooted in classical, Newtonian physics. The scientific revolutions in quantum physics, chaos theory, and complexity theory have demonstrated that the lockstep universal causal determinism assumed by scientific predeterminism does not exist. Rather, some indeterminism underlies physics, chemistry, and, especially, biology. Such indeterminism can, in certain circumstances, interrupt the operation of "laws" that have been generalized (including mathematically generalized) from complicated and nonuniform data. This is especially the case in biological venues such as the human brain, which is a combination of indeterministic "freedom" and deterministic "will" (deterministic in the sense that the will chooses among indeterministic thoughts and feelings and thereafter acts upon them in a causal manner).[1]

As biologist and complexity theorist Stuart A. Kauffman has observed, "no laws at all 'entail' the evolution of the biosphere, and . . . reductionism, [the] dream of a final [physics] theory, is false"[2] He elaborated on this point as follows:

> Pythagoras taught us that all is number. Newton, after Galileo, taught the same thing. Nature is written in number, the "rule and line" that Keats decried. It is a major transformation in our way of knowing the world if we can write no equations for the emergence of [biological organisms]. . . .
>
> This will be a major theme of the rest of this book. We cannot derive this becoming by equations. The becoming is not derivable by entailing law, for we can write no laws of motion for the evolving biosphere, as we do not know the relevant variables prior to their emergence in evolution . . . . We cannot mathematize the specific evolution of the biosphere. We can, at best, seek statistical laws about distributions of aspects of this evolution. In short, I will claim that no law at all entails the becoming of the biosphere; and that therefore, we cannot reduce biology to physics. The world is not a machine.[3]

Kauffman observed that "[w]ith the *Principia* of Newton, Aristotle's four causes—formal, final, efficient, and material—dwindled to a mathematized version of efficient cause: Newton's differential and integral calculus as captured in his three laws of motion and universal gravitation."[4]

In contrast, Aristotle had a much more comprehensive view of the meaning of cause. By what came to be known as "material cause," Aristotle meant "that out of which something comes into being, still being present in it, as the bronze of a statue or the silver of a bowl, or the kinds of these." He described what we call "formal cause" as "the form [*eidos*] or pattern [*paradeigma*], and this is the gathering in speech of what it is for something to be, or again the kinds of this (as of the octave, the two-to-one ratio, or generally number) and the parts that are in its articulation." What we call (mistakenly, according to translator Joe Sachs) "efficient cause" is "that from which the first beginning of change or rest is, as the legislator is a cause, or the father of a child, or generally the maker of what is made, or whatever makes a changing thing change." The "final cause" is "the end [*telos*]. This is that for the sake of which, as health is of walking around. Why is he walking around? We say 'in order to be healthy,' and in so saying think that we have completely given the cause."[5]

Modernity, starting with Francis Bacon and Thomas Hobbes, suppressed Aristotle's formal and final causes in favor of material and efficient causes.[6] The motivation was opposition to the medieval mindset wherein final cause (teleology) was construed in theological terms. As professional philosopher Cees

Leijenhorst has observed, "Hobbes's doctrine of causes may be seen as a systematic attempt to discard the scholastic view on causality and replace it with strict mechanistic explanatory principles."[7] Classical modern science has ardently desired to reduce everything to physical (material and efficient) causes. However, this project has proved to be impossible, because living organisms and their constituent parts obviously have external ends and internal functions as well as evolving species-defined forms. Biological beings, with self-preserving, self-fulfilling, and reproductive needs and strategies, are not rocks.[8]

Biology—including but not limited to human and other social animals—transforms the austere world of physics, refuting forever any notion that all natural (including mental) events operate according to presumed laws of scientific predeterminism. These are empirical facts *par excellence.*

So much for predeterminism. However, some commentators hedge and try to square the circle by claiming to be compatibilists.[9] We discussed compatibilism at the end of Chapter 1. Classical compatibilists/soft determinists hold that everything is predetermined but that we nevertheless have free will when no one physically interferes with our freedom. This is, as I argued above, self-contradiction and/or verbal trickery. If predeterminism is true, every effort to interfere with our freedom is predetermined and every response by us to such efforts is predetermined. As explained in the Introduction, free will is the independent ability to make conscious **decisions** that are neither predetermined nor random. Interference by others with one's physical freedom is not relevant to free will. Rather, such interference triggers ethical, legal, and political inquiries that presuppose free will.

Contemporary compatibilism has branched out in many directions, reminding one of nothing so much as the disputations of medieval Scholasticism. Compatibilists/soft determinists joust with free willists, scientific predeterminists, and each other with a jargonized vocabulary that has little or no application to real human life.

As discussed in Chapter 1, one problem with compatibilist analyses is that they are usually concerned with actions, not with choices and decisions. They lead to (or arise from) issues regarding moral and legal responsibility—questions that are subjects of ethical, legal, and political philosophy but hardly belong to the question of whether free will exists.

Today's compatibilists/soft determinists often posit imaginary "controllers" that allegedly disprove the idea that free will requires alternative possibilities.[10] Even assuming that a hypothetical "controller" could control thoughts (choices and decisions) as well as actions, this would have nothing to do with free will. As observed in the Introduction, brainwashed or hypnotized people obviously do not have free will: they do not have an "independent ability" to make "conscious decisions." The "controller" scenario may have been (and perhaps still is) a real thing in certain circumstances in some totalitarian countries (see, for example, the movie *The Manchurian Candidate*). But it has nothing to do with the lives of the overwhelming majority of people at any time in human history.

What warrants the certainty on the part of many (pre)determinists and compatibilists/soft determinists that free will (in the sense defined in the Introduction) does not exist? Thomas W. Clark, one of their number, has actually made a startling admission:

> Whether or not **naturalized conceptions of autonomy and agency** ever replace the **folk dualism of soul control**, and whatever the ultimate impact of understanding unconscious processes might be on our autonomy, **the subjective significance of phenomenal experience will stand unchallenged. And the puzzle of explaining consciousness, which matters considerably to scientists and philosophers, still stands as well. The first person reality of phenomenal experience, bringing with it the reality of moral concern, still awaits a conceptually and empirically transparent integration into a naturalized worldview.**[11]

Notice the false dichotomy between "soul control" (evoking religious dualism) and scientific predeterminism in the foregoing quotation. To the scientific (pre)determinists, it is a question of religion versus science, and they take the side of science in this centuries-long war. But, as we discussed in Chapter 2, there are secular scientists, especially neuroscientists, who argue that free will probably arose as a matter of biological evolution by natural selection. Can this position be absolutely proven at this moment in history? No. But then neither can scientific predeterminism, as indicated in the above quotation and in the preceding chapter of this book. Neither side can rightly claim final scientific victory. The difference is that most predeterminists opine with dogmatic certainty on the basis of assumptions that they cannot prove and that fail to account for all the empirical phenomena. In contrast, the scientists supporting free will speak, as we have shown in Chapter 2, with a tentativeness that befits the philosophic and scientific quest in the spirit of Socrates. "[W]hat I don't know," said Socrates, "I don't think I know."[12] The predeterminists claim they know what, in fact, they do not really know.

THE CASE FOR FREE WILL

I am not a scientist, but the argument by some scientists that free will probably arose as a matter of evolution by natural selection[13] makes a great deal of sense to me. I also find the neuroscientific analyses supporting free will of William R. (W. R.) Klemm, Peter Ulric Tse, Henry P. Stapp, and Jeffrey M. Schwartz, as discussed and referenced in Chapter 2, convincing. The overall two-stage framework of astrophysicist-philosopher Robert O. (Bob) Doyle, also discussed in that chapter, is cogent: indeterminism plays a role in the initial stage of thoughts or feelings presenting themselves to one's consciousness, followed by the agent making a choice among those possibilities and then acting on it insofar external circumstances permit such action. This two-stage model is, as I understand it, consistent with the views of Klemm, Stapp, and Schwartz. Tse has a three-stage model of free will which, however, does not differ significantly from Doyle's two-stage paradigm. Tse, a neuroscientist, develops in detail his own theory as to how exactly this process works itself out in the brain.

Stapp and Schwartz articulate a quantum theory of the mind, in which consciousness plays a decisive role. Physicist Ilya Prigogine and biologist Stuart A. Kaufmann are among those scientists who think it is important to go beyond both classical and quantum mechanics, as well as Einstein's relativity theory, to chaos and complexity theory.[14] Kauffman and professional philosopher Thomas Nagel have argued that natural selection, as narrowly construed by reductionist physicists, chemists, and biologists, is insufficient to establish free will. They propose to supplement Darwinian natural selection with additional paradigms. Kauffmann emphasizes the importance of self-organization of biological life in the context of complexity theory. Nagel appears to agree with Kauffman on this point and hypothesizes an overall teleological paradigm set in nature that may be metaphysical but is not theological (Nagel is a self-professed atheist).[15]

In the last analysis, all of these scholars (Doyle, Klemm, Tse, Stapp, Schwartz, Prigogine, Kauffman, Nagel, and others cited in the endnotes) argue that science, properly understood, supports the existence of free will. I myself do not have a sufficient background in physics, chemistry, biology, and mathematics to evaluate the details of the various scientific arguments for free will, but I think that the truth about free will resides somewhere within these interpretations or, perhaps, in some similar scientific perspective that will emerge in the future.

When I compare the arguments of the scientific predeterminists with the scientific and philosophical arguments of the proponents of free will, I conclude that free will does, in fact, exist. The next section discusses our subjective experience—what is called the "phenomenology"—of free will.

FREE WILL IN HUMAN LIFE

Let's start with simple things. We decide to get off of the sofa and do some exercise. Is this predetermined? Chapter 2 demonstrated that the conclusions of Libet, Wegner, and others claiming that that the unconscious mind initiates all such actions are scientifically and logically incorrect. Although we may experience conflicting and even unconscious impulses—to exercise or not to exercise—we consciously decide to do one or the other. Sure, we may be predisposed to exercise or, alternatively, to be a couch potato. Such predisposition is based on our past habits and our habitual mental frame of mind—our "characteristics," in Aristotle's terminology. But it is possible, in most cases, to change one's characteristics by a conscious determination to do so, followed by habituation in practicing the new lifestyle. Such deliberate mental training actually causes changes in the brain that make the new course of action a new habit. This involves the scientifically established neuroplasticity of the brain: see the subsections in Chapter 2 on Jeffrey M. Schwartz and Norman Doidge.

Even more simply, think about the decisions you make at every moment of your waking life, for example, deciding to reach for a cup of coffee, grab a fork, walk across a room, pick up a pen, type something on a laptop or desktop computer, call someone on a phone, text someone on a mobile phone, send an email, and so on and so forth. Are all these decisions and subsequent actions predetermined from the moment of the Big Bang? If so, we would be living absolutely ridiculous lives—playthings in the hands of the physics gods, as it were. When you think about it, each one of these simple decisions and ensuing actions, however automatic they may seem, once had to be learned in a conscious and deliberate way. For example, watching a baby learn to walk makes us realize how much of a conscious effort it was for us to have learned how to do that task. And, if you focus on it, you will realize that everything you do, absent a relevant medical condition, actually requires a preceding conscious decision on your part. Without an unusual medical condition or the use of physical force by another, your arm does not move unless you decide to move it. Think about this as you go about your daily activities, and you will see that at some level you are aware (even if not fully conscious) of the decisions that lead to your actions.

Then we come to the big life decisions. These questions start to confront us as early as high school. What are my plans after high school? Will I get a job and, if so, what kind of job? Will I go to a trade school? A community college? Will I apply for college and, if so, at what schools and how will I finance such an education? What are my long-term educational and work objectives? Shall I date a particular person? Should I get married and, if so, to whom? Once married, should my partner and I have children and, if so, how many?

Here is where the ad hoc determinists (see Chapter 1) chime in: It's all a matter of luck—genes and environment—how you will make these decisions. If you are born with good brains and in a wealthy, well-educated family, you might end up going to an Ivy League school, marrying a person with a matching upper-class background, getting a well-paying job in business or one of the professions, sending your kids to private school, etc. etc.

True. Life is not always fair. But each person, whatever their background, can work to improve their situation. We also see that people born with silver spoons in their mouths often get away with things the rest of us can't. This fact of life is all the more reason most of us need to think clearly and rationally about the choices we make. There is no "get out of jail free" card for the vast majority of us. We need to use our best judgment when confronted with choices.

As discussed in the Introduction, some people have severe medical conditions that affect their ability to exercise free will or to exercise it in a rational manner. These facts elicit our sympathy or empathy, and we do not hold such people to the same ethical standards that are rightly applied to most of us.

In the course of life, most of us are confronted with ethical issues or dilemmas. My next book, provisionally titled *Reason and Human Ethics*, will address such issues in depth. It is important to realize that, absent a serious medical problem, we have free will to think about and resolve ethical issues. It is a cop-out to claim that we are predetermined in our ethical decisions. We should face such questions as honestly and rationally as we can.

Finally, individual free will has political ramifications. The decisions and actions of political leaders are not predetermined. Whether or not the world is destroyed in a nuclear war will be the result of decisions made by political actors (whether voting citizens or governmental officials) exercising free will. The same goes for many other important political issues. My forthcoming book on ethics will address, among other things, citizen ethics and political ethics. I also plan to write, eventually, a work on political philosophy.

Humans may not have perfect free will, but we can normally exercise some degree of free will in most circumstances and work to improve our capacity for free will in other situations. We should not shrink from such opportunity. Rather, we should welcome it.

Conclusion

This book has demonstrated the following:
- The doctrines of divine predeterminism, scientific predeterminism, and ad hoc determinism are all based on logical and scientific errors.
- The doctrine of compatibilism/soft determinism, in its typical formulation, requires its proponents to commit a violation of the principle of (non)contradiction and, accordingly, to entertain cognitive dissonance of a high order. Alternatively, it changes the traditional definitions of such words as "free will" and "(pre)determinism" to mean something different from what is normally meant by those terms in philosophical and scientific discourse, thereby rendering the discussion irrelevant.
- "Free will" is the independent ability to make conscious decisions that are neither predetermined nor random. There are some arguments for free will that are persuasive and that justify further research and study. The first is that free will exists as a scientific fact, based on evolutionary and emergent developments in living beings. Unlike inorganic things, biological life is not subject to blind mechanism but is, to the contrary, teleological (end-seeking) in a

nonreligious sense. The second argument is that free will is established by quantum physics. This is more difficult for the layperson to understand and is dependent on such proven but unintuitive concepts such as quantum entanglement. Finally, the advocates of chaos and complexity theory have made a strong case for their approach to free will. Perhaps, in the last analysis, all these viewpoints can and will be merged. However this may be, human experience supports the principle that free will, in some form, does exist, and this conclusion is supported by contemporary science.

All is not lost. The future is not predetermined but rather affords us possibilities for our individual and collective good or ill. We humans, both individually and collectively, determine our own ethical, social, and political future through our exercise of free will. The choice is up to us. By nature, we cannot escape that burden. This is our distinction as human beings.

NOTES

Introduction

[1] William Shakespeare, *Macbeth*, 5.5.2345–49, in *The First Folio of Shakespeare: The Norton Facsimile*, 2nd ed., ed. Charles Hinman and Peter W. M. Blayney (New York: W.W. Norton & Co., 1996), 758 (line breaks omitted, capitalization and spelling modified).

[2] Pierre Simon Laplace, *A Philosophical Essay on Probabilities*, trans. (from the 6th French ed. published in 1814) Frederick Wilson Truscott and Frederick Lincoln Emory (New York: John Wiley and Sons, 1902), 3–4; Robert Kane, *A Contemporary Introduction to Free Will* (New York: Oxford University Press, 2005), 7–8; Peter Ulric Tse, *The Neural Basis of Free Will: Criterial Causation* (Cambridge, MA: MIT Press, 2013), §§ A1.1–A1.3; Gregg D. Caruso, *Free Will and Consciousness: A Determinist Account of the Illusion of Free Will* (Lanham, MD: Lexington Books, 2012), 2, 8–9, 12–13n1, 14n8; Gregg D. Caruso, "Introduction," in *Exploring the Illusion of Free Will and Moral Responsibility*, ed. Gregg D. Caruso (Lanham, MD: Lexington Books, 2013), 15n4, Kindle; Christian List, *Why Free Will is Real* (Cambridge, MA: Harvard University Press, 2019), 38–43, 79–80; Henry P. Stapp, *Quantum Theory and Free Will: How Mental Intentions Translate into Bodily Actions* (Cham, SZ: Springer, 2017), ix, 61.

[3] Derk Pereboom, *Free Will, Agency, and Meaning in Life* (Oxford: Oxford University Press, 2014), 1, Kindle; Caruso, *Free Will and Consciousness*, 13n1; Caruso, "Introduction," 15n4.

⁴ Kane, *Contemporary Introduction*, 8–9; Bob (Robert O.) Doyle, *Free Will: The Scandal in Philosophy* (Cambridge, MA: I-Phi Press, 2011), 223–24, Kindle; Bob (Robert O.) Doyle, "Emergent Determinism," The Information Philosopher, accessed June 26, 2021, http://informationphilosopher.com/freedom/emergent_determinism.html; Tse, *Neural Basis of Free Will*, §§ A1.4–A1.7, A2.1; Manuel Vargas, "If Free Will Does Not Exist, Neither Does Water," in Caruso, *Illusion of Free Will*, 182.

⁵ Kane, *Contemporary Introduction*, 9; Caruso, *Free Will and Consciousness*, 13n3, 15–16, 18, 21; Galen Strawson, "The Impossibility of Ultimate Responsibility?," in Caruso, *Illusion of Free Will*, 49–50; Susan Blackmore, "Living without Free Will," in Caruso, *Illusion of Free Will*, 162; Ted Honderich, *How Free Are You: The Determinism Problem*, 2nd ed. (Oxford: Oxford University Press, 2002), 44.

⁶ Regarding the fallacy of false dichotomy or false dilemma, see, among other sources, Jennifer Culver, "False Dilemma," in *Bad Arguments: 100 of the Most Important Fallacies in Western Philosophy*, ed. Robert Arp, Steven Barbone, and Michael Bruce (Hoboken, NJ: Wiley-Walton, 2018), chap. 81, Kindle; W. Ward Fearnside and William B. Holther, *Fallacy: The Counterfeit of Argument* (Englewood Cliffs, NJ: Spectrum, 1959), 32–33; Douglas Walton, *Informal Logic: A Pragmatic Approach*, 2nd ed. (New York: Cambridge University Press, 2008), 52–53, 259, Kindle; Michael C. LaBossiere, *76 Fallacies* (Amazon Digital Services, 2012), 68–69, Kindle.

⁷ See Tse, *Neural Basis of Free Will*, § 1.3.

⁸ E.g., Caruso, "Introduction," 4, 5, 9, 13; Bruce Waller, "The Stubborn Illusion of Moral Responsibility," in Caruso, *Illusion of Free Will*, 65, 68, 71, 72, 75; Saul Smilanksy, "Free Will as a Case of 'Crazy' Ethics," in Caruso, *Illusion of Free Will*, 115. Neuroscientist Peter Tse asks: "Is the

commonsense point of view, that, for example, we go to the dentist because a tooth hurts, an illusion or delusion as reductionistic materialists typically imply when they dismiss this view as mere 'folk psychology?' " Tse, *Neural Basis of Free Will*, § 1.3. He suggests that "as neuroscience matures, and outgrows the insecurities that led to behaviorism's rejection of experience as a valid domain of scientific inquiry, I suspect we will see the emergence of a 'Gestalt Neuroscience' " that includes experience as a proper object of scientific study. Tse, § 1.12.

[9] Plato, *Apology of Socrates* 20c–23b. For examples, see Plato's Socratic dialogues.

[10] Doyle, *Free Will*, 252, 261, 325, 368, 369, 381–82; Robert O. Doyle, "The Two-Stage Model to the Problem of Free Will: How Behavioral Freedom in Lower Animals Has Evolved to Become Free Will in Humans and Higher Animals," in *Is Science Compatible with Free Will?: Exploring Free Will and Consciousness in the Light of Quantum Physics and Neuroscience*, ed. Antoine Suarez and Peter Adams (New York: Springer Science+Business Media, 2013), 235–36, 239–41, 252; Tse, *Neural Basis of Free Will*, §§ 3.8. 4.5, 6.22, 7.11, 10.7–10.11, 10.45–10.47; Peter Ulric Tse, "Two Types of Libertarian Free Will Are Realized in the Human Brain," in *Neuroexistentialism: Meaning, Morals, and Purpose in the Age of Neuroscience*, ed. Gregg D. Caruso and Owen Flanagan (New York: Oxford University Press, 2018), 181, 184, 189; Helen Steward, *A Metaphysics for Freedom* (Oxford: Oxford University Press, 2012), esp. chap. 4. Cf. Antonio Damasio, *Descartes' Error: Emotion, Reason, and the Human Brain* (New York: Penguin Books, 1994), 89–90, Kindle.

[11] Doyle, *Free Will*, 127.

¹² In addition to Doyle's book *Free Will*, see the discussions on his website at http://informationphilosopher.com/freedom/.html.

¹³ Doyle, *Two-Stage Model*, 243. For what Doyle means by "adequate determinism," see the discussion on his website at https://informationphilosopher.com/freedom/adequate_determinism.html.

¹⁴ William R. (W. R.) Klemm, *Making a Scientific Case for Conscious Agency and Free Will* (n.p.: Elsevier, Academic Press, 2016), 3 (italics omitted), Kindle.

¹⁵ Aristotle, *Nicomachean Ethics*, 1114b–15a (see also the section on Aristotle in Chapter 2 of this book); W. R. Klemm, "Free Will Debates: Simple Experiments Are Not So Simple," *Advances in Cognitive Psychology* 6 (2010): 48, 55, 56, 61, https://doi.org/10.2478/v10053-008-0076-2; Susan Pockett, "The Neuroscience of Movement," in *Does Consciousness Cause Behavior?*, ed. Susan Pockett, William P. Banks, and Shaun Gallagher (Cambridge, MA: MIT Press, 2006), 19–20; John-Dylan Haynes and Michael Pauen, "The Complex Network of Intentions," in Caruso, *Illusion of Free Will*, 226–27, 233. Cf. Tse, *Neural Basis of Free Will*, §§ 2.11, 3.3n3, 9.7–9.11, 10:53–10.56, 10.58–10.59, 10.80; Tse, "Two Types of Libertarian Free Will," 182; Damasio, *Descartes' Error*, 166–67; Robert Hanna and Michelle Maiese, *Embodied Minds in Action* (Oxford: Oxford University Press, 2009), 91; Alfred R. Mele, *Effective Intentions: The Power of Conscious Will* (New York: Oxford University Press, 2009), 36–38, Kindle; Andrea Lavazza, "Why Cognitive Sciences Do Not Prove That Free Will Is an Epiphenomenon," *Frontiers in Psychology* 10, no. 326 (2019): 9, https://www.academia.edu/38462524/Why_Cognitive_Sciences_Do_Not_Prove_That_Free_Will_Is_an_Epiphenomenon?email_work_card=view-paper.

[16] Marc Jeannerod, "Consciousness of Action as an Embodied Consciousness," in *Does Consciousness Cause Behavior?*, ed. Pockett, Banks, and Gallagher, 25.

[17] See Michael Schirber, "The Chemistry of Life: The Human Body," Live Science, April 16, 2009, https://www.livescience.com/3505-chemistry-life-human-body.html; Tse, *Neural Basis of Free Will*, § 9.7; Hanna and Maiese, *Embodied Minds in Action*, 41; and Mele, *Effective Intentions*, 86 ("we are not conscious of . . . neural events as neural events"), Kindle.

[18] Consider the type of scenario that is presented in the movie *The Manchurian Candidate*. Similarly, neuroscientist Peter Ulric Tse observes: "Many people who act under posthypnotic suggestions or who suffer from alien hand syndrome do not act freely or from free will, arguably because they are not aware of any choice and form no conscious intention to act . . . ." Tse, *Neural Basis of Free Will*, § 3.3n3; cf. Kane, *Contemporary Introduction*, 15.

[19] Doyle, *Free Will*, 417.

[20] Carl Hoefer, "Causal Determinism," The Stanford Encyclopedia of Philosophy, spring 2016 ed., ed. Edward N. Zalta, first sentence, accessed June 26, 2021, https://plato.stanford.edu/entries/determinism-causal/.

[21] Doyle, *Free Will*, 145; see also Caruso, "Introduction," 2.

[22] Cf. Tse, *Neural Basis of Free Will*, §§ 3.1, 3.3, 3.3n2, 4.16; and Robert Hanna, *Deep Freedom and Real Persons—A Study in Metaphysics* (New York: Nova Science, 2018), 27 (vol. 2 of Hanna, *The Rational Human Condition*).

Chapter 1. Arguments against Free Will

[1] Stephen Hawking, "Is Everything Determined?," in *Black Holes and Baby Universes and Other Essays* (New York: Bantam Books, 1994), 134.

[2] Caruso, "Introduction," 23.

[3] For a philosophical analysis of theological predeterminism (with a focus on Christian predeterminism), see William Hasker, "Divine Knowledge and Human Freedom," in *The Oxford Handbook of Free Will*, 2nd ed., ed. Robert Kane (New York: Oxford University Press, 2011), 39–54. The history of Christian predeterminism is discussed in some detail in Jesse Couenhoven, *Predestination: A Guide for the Perplexed* (London: T&T Clark, 2018), Kindle. I express no opinion regarding the accuracy of these accounts and interpretations.

[4] Romans 8:29–30, New Revised Standard Version (NRSV), in *HarperCollins Study Bible*, rev. ed., ed. Harold W. Attridge and Wayne A. Meeks (New York: HarperCollins, 2006), Kindle.

[5] Romans 8:33 (NRSV).

[6] Augustine, *On Free Choice of the Will*, trans. Thomas Williams (Indianapolis: Hackett, 1993), Kindle.

[7] Michael Mendelson, "Saint Augustine," The Stanford Encyclopedia of Philosophy, winter 2018 ed., ed. Edward N. Zalta, §§ 1, 7, accessed May 29, 2019, https://plato.stanford.edu/archives/win2018/entries/augustine.

[8] Robert Van de Weyer, "Introduction," in *The Letters of Pelagius: Celtic Soul Friend*, ed. Robert Van de Weyer (Worcestershire, UK: Arthur James Ltd., 1995); Augustine, *Reconsiderations*, I.9.5, in *Free Choice of the Will*, 128.

[9] Augustine, *Reconsiderations*, bk. 1, chap. 9, in *Free Choice of the Will*, 124–29.

[10] Augustine, *Reconsiderations*, bk. 1, chap. 9, § 4, 127.

[11] Augustine, *A Treatise on the Predestination of the Saints*, trans. Philip Schaff, in Augustine, *The Collected Works of 26 Essential Treatises*, ed. Philip Schaff (Amazon Digital Services, 2011), loc. 17403–18530 of 44835, Kindle.

[12] Augustine, *Predestination*, chap. 7, loc. 17544–46.
[13] Augustine, chap. 16, loc. 17857–59.
[14] Augustine, chap. 16, loc. 17859–63.
[15] Augustine, chap. 19 [X], loc. 17918–21.
[16] Augustine, chap. 34 [XVII], loc. 18300.
[17] Augustine, chap. 34 [XVII], loc. 18306–7.
[18] Augustine, chap. 35 [XVIII], loc. 18364–65.
[19] Martin Luther, *The Bondage of the Will* (1525), trans. J. I. Packer and O. R. Johnston (Grand Rapids, MI: Fleming H. Revell, 1957).
[20] Martin Luther, "Preface to the Epistle of St. Paul to the Romans" (1522), trans. Bertram Lee Woolf, in *Martin Luther: Selections from His Writings*, ed. John Dillenberger (New York: Anchor Books, 1962), 32.
[21] Luther, *Bondage of the Will*, 80 (emphasis added).
[22] Luther, 83 (emphasis added).
[23] John Calvin, *Institutes of the Christian Religion*, trans. Henry Beveridge (1845; repr., Grand Rapids, MI: Wm. B. Eerdmans, 2001), bk. 3, chap. 21, § 4.
[24] John Calvin, "Articles concerning Predestination," in *Calvin: Theological Treatises*, trans. and ed. J. K. S. Reid (Philadelphia: Westminster, 1954), 179.
[25] Calvin, *Institutes*, bk. 3, chap. 23, § 6.
[26] Thomas Hobbes, *Leviathan*, rev. student ed., ed. Richard Tuck (Cambridge: Cambridge University Press, 1996), pt. 3, esp. chaps. 32, 35, 37, 38, 39, 42, and 43 passim.
[27] Leo Strauss, *The Political Philosophy of Hobbes: Its Basis and Genesis*, trans. Elsa M. Sinclair (Chicago: University of Chicago Press, 1952), 74.
[28] Strauss, 76–77.
[29] *Hobbes and Bramhall on Liberty and Necessity*, ed. Vere Chappell (Cambridge: Cambridge University Press, 1999).

[30] Chappell, introduction to *Hobbes and Bramhall*, 6.
[31] Hobbes, *The Questions concerning Liberty, Necessity, and Chance* (published 1656), § 20(n), in Chappell, *Hobbes and Bramhall*, 80.
[32] Luther, *Bondage of the Will*, 80.
[33] Calvin, *Institutes*, bk. 3, chap. 23, § 6.
[34] Hobbes, *Of Liberty and Necessity* (published 1654), § 11, in Chappell, *Hobbes and Bramhall*, 20 (emphasis added).
[35] Hobbes, § 11, p. 21.
[36] Hobbes, § 17, p. 29 (emphasis added).
[37] Hobbes, § 35, p. 41.
[38] Hobbes, § 36, p. 41 (emphasis added).
[39] Hobbes, *Questions concerning Liberty, Necessity, and Chance*, "To the reader," 69 (emphasis added).
[40] E.g., Hobbes, *Of Liberty and Necessity*, § 12, 21–23.
[41] Hobbes, *Questions concerning Liberty, Necessity, and Chance*, "The occasion of the controversy," 70.
[42] Hobbes, *Of Liberty and Necessity*, § 24, p. 35 (emphasis added).
[43] "I do indeed take all voluntary acts to be free, and all free acts to be voluntary; but withal that all acts, whether free or voluntary, if they be acts, were necessary before they were acts." Hobbes, *Questions concerning Liberty, Necessity, and Chance*, § 28(a), pp. 82–83.
[44] Conflicting claims to revelation include, for example, Hinduism, the oracle of the ancient Egyptian god Amon, Judaism, Zoroastrianism, the ancient Greek Homeric gods and Delphic oracle, Norse mythology, Christianity, Gnosticism, Manicheism, Islam, Sikhism, Swedenborgism, the Shakers (revelation to Mother Anne Lee), the Bahá'í faith, the Latter Day Saints (revelation to Joseph Smith), the revelation to the Paiute Native American prophet Wovoka (Jack Wilson), Thelema (revelation to Aleister Crowley), and televangelists

Oral Roberts and Pat Robertson (who both claimed to speak with God).

[45] Immanuel Kant, *Lectures on the Philosophical Doctrine of Religion*, trans. Allen W. Wood, in Kant, *Religion and Rational Theology*, ed. Allen W. Wood and George di Giovanni (New York: Cambridge University Press, 1996), 28:1115–16 (German Academy pagination, italics in the original, editorial note omitted).

[46] Doyle, *Free Will*, 3.

[47] Laplace, *Philosophical Essay on Probabilities*, 3–4.

[48] List, *Why Free Will Is Free*, 80.

[49] Ted Honderich, *How Free Are You: The Determinism Problem*, 2nd ed. (Oxford: Oxford University Press, 2002), 7.

[50] See also Ted Honderich, "Determinism, Compatibilism and Incompatibilism, Actual Consciousness and Subjective Physical Worlds, Humanity" in Caruso, *Exploring the Illusion of Free Will*, 54.

[51] See also Honderich, "Determinism, Compatibilism," 55; Ted Honderich, "Effects, Determinism, Neither Compatibilism nor Incompatibilism, Consciousness" (2011), in Kane, *Oxford Handbook of Free Will*, 2nd ed., 446–48.

[52] Stapp, *Quantum Theory and Free Will*, 13.

[53] See Wayne R. LaFave and Austin W. Scott Jr., *Criminal Law* (St. Paul, MN: West, 1972), 24.

[54] "[M]any scientists and philosophers have used Libet's and similar results to argue against the existence of free will and responsibility . . . ." Tse, *Neural Basis of Free Will*, § 9.2. "Although Libet himself stopped short of endorsing free will scepticism on the basis of his results, other theorists have not been so cautious, and his work is often said to show that we lack free will." Tim Bayne, "Libet and the Case for Free Will Scepticism," in *Free Will and Modern Science*, ed. Richard Swinburne (New York: Oxford University Press, 2011), 26,

Kindle. See, for example, the sections on Daniel Wegner and Sam Harris later in the present chapter.

[55] Benjamin Libet, *Mind Time: The Temporal Factor in Consciousness* (Cambridge, MA: Harvard University Press, 2004).

[56] See generally Klemm, *Scientific Case for Free Will*, 8–10, 25; List, *Why Free Will Is Real*, 141–47; Lavazza, "Cognitive Sciences Do Not Prove That Free Will Is an Epiphenomenon," 5.

[57] Klemm, "Free Will Debates," 48, 50, 54–57, 61; Klemm, *Scientific Case for Free Will*, 25; Tse, *Neural Basis of Free Will*, § 9.9–9.11; Alfred R. Mele, *Free: Why Science Hasn't Disproved Free Will* (New York: Oxford University Press), 13–14, Kindle; Mele, *Effective Intentions*, 53–55 and chap. 6; cf. Walter J. Freeman, *How Brains Make Up Their Minds* (New York: Columbia University Press, 2000), 123–24.

[58] Klemm, "Free Will Debates," 50; W. R. (William R.) Klemm, *Atoms of Mind: The "Ghost in the Machine" Materializes* (n.p.: Springer Science+Business Media, 2011), 263, Kindle.

[59] Klemm, "Free Will Debates," 50.

[60] Klemm, 55.

[61] Klemm, 50; Bob Doyle, "Benjamin Libet," The Information Philosopher, accessed May 3, 2021, https://informationphilosopher.com/solutions/scientists/libet/; Tse, *Neural Basis of Free Will*, §§ 9.2, 9.5; Mark Balaguer, *Free Will* (Cambridge, MA: MIT Press, 2014), 98–101, Kindle; Mele, *Free*, 12–13, 18–22; Eddy Nahmias, "Free Will and Responsibility," Wiley Interdisciplinary Reviews: Cognitive Science (2012), https://www.academia.edu/41577478/Free_will_and_responsibility; Lavazza, "Cognitive Sciences Do Not Prove That Free Will Is an Epiphenomenon," 5.

[62] Jeff Miller, Peter Shepherdson, and Judy Trevena, "Effects of Clock Monitoring on Electroencephalographic Activity: Is Unconscious Movement Initiation an Artifact of the Clock?," *Psychological Science* 22, no. 1 (January 2011): 103–4, https://www.jstor.org/stable/40984614.

[63] Tse, *Neural Basis of Free Will*, § 9.2. See also Mele, *Effective Intentions*, chaps. 3–4 passim.

[64] Stapp, *Quantum Theory and Free Will*, 67 (italics in the original); see also Henry P. Stapp, "Quantum Theory of Mind," in *Contemporary Dualism: A Defense*, ed. Andrea Lavazza and Howard Robinson (New York: Routledge, 2014), 109–111, Kindle.

[65] Mele, *Effective Intentions*, chap. 4.

[66] Klemm, "Free Will Debates," 55, 58–59; Doyle, "Benjamin Libet"; Mele, *Effective Intentions*, 85; Lavazza, "Cognitive Sciences Do Not Prove That Free Will Is an Epiphenomenon," 5.

[67] Klemm, "Free Will Debates," 55; cf. Tse, *Neural Basis of Free Will*, §§ 9.7–9.8.

[68] Klemm, "Free Will Debates," 55.

[69] See, e.g., Michael J. Muniz, "Hasty Generalization," in Arp, Barbone, and Bruce, *Bad Arguments*, chap. 84; Walton, *Informal Logic*, 2nd ed., 162–63, 246; Ingo Brigandt and Alan Love, "Reductionism in Biology," The Stanford Encyclopedia of Philosophy, spring 2017 ed., ed. Edward N. Zalta, https://plato.stanford.edu/archives/spr2017/entries/reduction-biology/; Klemm, "Free Will Debates," 58–60; Tse, *Neural Basis of Free Will*, § 9.14; Mele, *Free*, 13–17; Nahmias, "Free Will and Responsibility."

[70] Doyle, "Benjamin Libet"; Stapp, *Quantum Theory and Free Will*, 67. Cf. Klemm, *Scientific Case for Free Will*, 30–31; Nahmias, "Free Will and Responsibility" (the initial readiness potential may represent only an "urge" rather than a fully formed intention or decision to flex).

⁷¹ Klemm, "Free Will Debates," 55–60 passim; Klemm, *Scientific Case for Free Will*, 8–12, 25, 57–58, 72; Tse, *Neural Basis of Free Will*, § 9.14; Bob Doyle, "John-Dylan Haynes," The Information Philosopher, accessed April 24, 2021, https://informationphilosopher.com/solutions/scientists/haynes/; Balaguer, *Free Will*, 101–17; Nahmias, "Free Will and Responsibility"; Mele, *Free*, chap. 3; Mele, *Effective Intentions*, chap. 6 passim; Lavazza, "Cognitive Sciences Do Not Prove That Free Will Is an Epiphenomenon," 5.

⁷² The edition used in the present book is Daniel M. Wegner, *The Illusion of Conscious Will*, New Edition (Boston: MIT Press, 2018), Kindle. This Kindle edition does not contain the print book's page numbers.

⁷³ Wegner, *Illusion of Conscious Will*, chap. 2 passim.

⁷⁴ Libet, *Mind Time*, 155–56; see also 144, 152.

⁷⁵ Wegner, *Illusion of Conscious Will*, chaps. 1–2, 4–8 passim.

⁷⁶ Mele, *Free*, chap. 4; Nahmias, "Free Will and Responsibility"; Tse, *Neural Basis of Free Will*, § 9.14; Klemm, *Scientific Case for Free Will*, 10–11.

⁷⁷ Wegner, *Illusion of Conscious Will*, chaps. 3–4 passim.

⁷⁸ Wegner, chap. 2, loc. 1296 of 8334, Kindle.

⁷⁹ Alfred Einstein, *Mozart: His Character, His Work*, trans. Arthur Mendel and Nathan Broder (New York: Oxford University Press, 1945), 24–25.

⁸⁰ Wegner's discussion and examples of this topic are in chapter 3 of his *Illusion of Conscious Will*.

⁸¹ For additional critiques of Wegner's position, see Mele, *Effective Intentions*, chaps. 2 and 5; Tim Bayne, "Phenomenology and the Feeling of Doing: Wegner on the Conscious Will," in Pockett, Banks, and Gallagher, *Does Consciousness Cause Behavior?*, 169–85; Klemm, *Scientific Case for Free Will*, § 3.4; Klemm, *Atoms of Mind*, 256; Lavazza, "Cognitive Sciences Do Not Prove That Free Will Is an Epiphenomenon," 6.

[82] Leo Strauss, *Persecution and the Art of Writing* (Chicago: University of Chicago Press, 1952), 30.

[83] Sam Harris, *Free Will* (New York: Free Press, 2012), 5 (italics in the original), Kindle.

[84] Harris, 39 (italics in the original).

[85] Susan Blackmore, "Living without Free Will," in Caruso, *Illusion of Free Will*, 161–76.

[86] Blackmore, 163.

[87] Blackmore, 164–65.

[88] F. Scott Fitzgerald, *The Great Gatsby* (1925; repr. with corrections, New York: Scribner Paperback Fiction, 1992), 5–6.

[89] Neil Levy, "Be a Skeptic, Not a Metaskeptic," in Caruso, *Exploring the Illusion of Free Will*, 87.

[90] Klemm, *Scientific Case for Free Will*, 3; see also Kane, *Contemporary Introduction*, 10.

[91] Caruso, *Free Will and Consciousness*, 13n3.

[92] Caruso, 13n3.

[93] Caruso, 4, 8–9.

[94] List, *Why Free Will Is Real*, 98 (italics in the original).

[95] Damasio, *Descartes' Error*, 176–77. See also Lavazza, "Cognitive Sciences Do Not Prove That Free Will Is an Epiphenomenon," 7–9.

[96] See, for example, the discussions of the experiments and analyses of Benjamin Libet and Daniel Wegner, discussed earlier in this chapter.

[97] Thomas W. Clark, "Experience and Autonomy: Why Consciousness Does and Does Not Matter," in Caruso, *Exploring the Illusion of Free Will*, 247; see also his elaboration of this point at 245–47. Cf. Hanna, *Deep Freedom and Real Persons*, 82–86.

[98] Michael McKenna, "Compatibilism," The Stanford Encyclopedia of Philosophy, spring 2021 ed., ed. Edward N.

Zalta, second sentence, accessed April 9, 2021, https://plato.stanford.edu/archives/spr2021/entries/compatibilism/. McKenna uses the term "determinism," which, in context, means what the Introduction to the present book defines as "predeterminism."

[99] See Hanna, *Deep Freedom and Real Persons*, § 1.2; List, *Why Free Will Is Real*, 151–52. Cf. Bob Doyle, "Comprehensive Compatibilism," The Information Philosopher, accessed June 12, 2021, https://informationphilosopher.com/freedom/comprehensive_compatibilism.html.

[100] Hobbes, *Questions concerning Liberty, Necessity, and Chance*, § 28(a), pp. 82–83; see also McKenna, "Compatibilism," § 2.1.

[101] Immanuel Kant, *Critique of Practical Reason* (1788), 5:95–97. See the Bibliography for the translation and edition of Kant's *Critique of Practical Reason* used herein.

[102] William James, "The Dilemma of Determinism," in *The Will to Believe: And Other Essays in Popular Philosophy* (New York: Longmans, Green, 1912), 68, Kindle.

[103] David Hume, *An Enquiry Concerning Human Understanding*, § 8, esp. ¶ 23, in *Hume: An Enquiry Concerning Human Understanding and Other Writings*, ed. Stephen Buckle (Cambridge: Cambridge University Press, 2007), 85, Kindle; see also McKenna, "Compatibilism," § 2.2. Hume was somewhat more explicit in his earlier work *A Treatise of Human Nature*, bk. 2, pt. 3, §§ 1–3, but in an "Advertisement" prefaced to *An Enquiry Concerning Human Understanding*, he called the *Treatise* a "juvenile work, which the author never acknowledged" (it was published anonymously). He concluded the "Advertisement" by requesting that the *Enquiry Concerning Human Understanding* "alone be regarded as containing his philosophical sentiments and principles" about these matters.

Hume expressed similar disparaging sentiments regarding the *Treatise* in a February 1754 letter to John Stewart (above-cited Buckle edition of *An Enquiry Concerning Human Understanding* at 210).

[104] Daniel C. Dennett, *Freedom Evolves* (New York: Penguin, 2004), 98, Kindle.

[105] My thanks to independent philosopher Robert Hanna for helping me clarify the thought expressed in this sentence, including actually supplying the language for it.

[106] Dennett, *Freedom Evolves*, 98 (italics in the original, terminal parenthesis marks omitted).

[107] Daniel C. Dennett, *Elbow Room: The Varieties of Free Will Worth Having*, New Edition (Cambridge, MA: MIT Press, 2015), 80 (italics in the original). The first edition of this book was published in 1984.

[108] Dennett, 136.

[109] Dennett, 135–37.

[110] Dennett, 188–206 passim.

[111] Dennett, ix. The concepts that Dennett considers outdated are what he calls "libertarian freedom" and "agent causation." These are technical terms used in the scholarly literature on free will. The word "libertarian" in this context has a philosophical meaning; it does not refer to political libertarianism. Philosophical libertarianism is "the view that (1) free will and determinism are incompatible (incompatibilism), (2) free will exists, and so (3) determinism is false." Kane, *Contemporary Introduction*, 32–33. The term "agent causation" is more complicated, because there have been different views of what constitutes agent causation. My approach is similar to what Aristotle meant when he wrote that that certain things are "up to us" (see the first section of Chapter 2; see also the section "Contemporary Philosophical and Scientific Arguments for Free Will" later in that chapter and Chapter 3).

[112] Dennett, *Elbow Room*, x–xi.

[113] Daniel C. Dennett, *Brainstorms: Philosophical Essays on Mind and Psychology*, 40th anniversary ed. (Cambridge, MA: MIT Press, 2017) (originally published in 1978), Kindle.

[114] Dennett, 396 (emphasis added).

[115] Dennett, 400.

[116] Jennifer Schuessler, "Philosophy That Stirs the Waters," *New York Times*, April 29, 2013, https://www.nytimes.com/2013/04/30/books/daniel-dennett-author-of-intuition-pumps-and-other-tools-for-thinking.html?searchResultPosition=1.

[117] Dennett, *Brainstorms*, 399–401.

[118] Dennett, 17 (emphasis added); cf. Dennett, *Freedom Evolves*, 126.

[119] McKenna, "Compatibilism," §§ 3–4; see also the discussion of compatibilism/soft determinism in Chapter 3 below.

Chapter 2. Arguments for Free Will

[1] William Shakespeare, *Julius Caesar*, act 1, lines 238–40, in *The First Folio of Shakespeare*, 2nd ed., 719.

[2] All quotations of Aristotle's *Nicomachean Ethics* are from *Aristotle's "Nicomachean Ethics,"* trans. and ed., Robert C. Bartlett and Susan D. Collins (Chicago: University of Chicago Press, 2011). The standard Bekker pagination (as denoted in this translation) is set forth parenthetically in the text. Bekker line references are approximate.

[3] Cf. Aristotle, *Eudemian Ethics* 1226b2–37.

[4] "It is not possible for the same thing at the same time both to belong and not belong to the same thing in the same respect . . . ." Aristotle, *Metaphysics* 1005b19–21, in *Aristotle's "Metaphysics"*, trans. Joe Sachs (Santa Fe, NM: Green Lion Press, 2002), 59; see also 1005b7–35, 1061b36–

1062a19. Of course, as discussed in the preceding chapter, many contemporary compatibilists, redefine such terms as "free will" and "(pre)determinism" in order to avoid the contradiction. (Such recently discovered phenomena as quantum mechanics raise questions whether the principle of contradiction applies at the frontiers of physics, but any such exception would not apply to this issue.)

[5] Aristotle then states that "not the end, but rather the things conducive to the end, would be the object of deliberation." It is strange to me that Aristotle would not consider deliberation appropriate for ends as well as means. In an "Interpretive Essay" affixed to their translation of the *Nicomachean Ethics*, Robert C. Bartlett and Susan D. Collins indicate the problematic character of this position and seem to suggest that it might be provisional in light of what comes before and what comes after in the work. *Aristotle's "Nicomachean Ethics"*, 255–56.

[6] Aristotle characteristically uses conditional clauses and hedging words like "seem" in his writing. Accordingly, it is sometimes difficult to ascertain exactly his own views. He was, however, more definite on this point in the *Eudemian Ethics*, where he wrote that humans are (when acting voluntarily rather than involuntarily, i.e., when it is up to us) "the principle/starting point/origin" and "controlling authority" of their actions. Such actions do not occur by necessity, chance, or nature but rather are up to us. Accordingly virtue and vice are voluntary. *Eudemian Ethics* 1223a3–20 (my translation of the quoted material).

[7] For a scholarly engagement with Aristotle's views expressed in the present section of this book, including but not limited to the passage just quoted, see Richard Sorabji, *Necessity Cause and Blame: Perspectives on Aristotle's Theory* (London: Duckworth, 1983), 233–38. Sorabji agrees with an

interpretation of Aristotle that emphasizes the latter's opposition to (pre)determinism.

[8] Bartlett and Collins, *Aristotle's "Nicomachean Ethics"*, 306 (glossary entry for "characteristic").

[9] As the Introduction to this book acknowledges, there are a small percentage of people who are born with genetic defects that affect their mental and ethical behavior.

[10] Robert Kane, *The Significance of Free Will* (New York: Oxford University Press, 1998): see his index entries for "self-forming actions" and "self-forming willings." Ebook readers can conduct word searches for same in the ebook edition (note: the endnotes in the Kindle edition were not hyperlinked at the time I downloaded the Kindle edition in 2020).

[11] Kane, 220n3, quoting Aristotle, *Physics* 256a6–8 (evidently Kane's translation).

[12] René Descartes, "Early Writings: Preliminaries," in *The Philosophical Writings of Descartes*, trans. John Cottingham, Robert Stoothoff, and Dugald Murdoch (New York: Cambridge University Press, 1985), 1:2, Kindle.

[13] René Descartes to Marin Mersenne, April 1634, in *The Philosophical Writings of Descartes*, trans. John Cottingham et al. (New York: Cambridge University Press, 1991), 3:42–44, Kindle; René Descartes, *Discourse on Method*, trans. Richard Kennington, ed. Pamela Kraus and Frank Hunt (Newburyport, MA: Focus, 2007), 48 (*Discourse*, pt. 6, near beginning); cf. 38 (pt. 5); see also Richard Kennington, "Interpretive Essay: Descartes's *Discourse on Method*," in Descartes, *Discourse on Method*, 59–60.

[14] Arthur M. Melzer, *Philosophy Between the Lines: The Lost History of Esoteric Writing* (Chicago: University of Chicago Press, 2014), 140, Kindle (citing Descartes to Mersenne, April 1634, in *Œuvres de Descartes*, 1:284–91; Ovid, *Tristia* 3.4.25).

[15] Damasio, *Descartes' Error*, 249.

[16] Richard Kennington, "René Descartes," in *History of Political Philosophy*, 3rd ed., ed. Leo Strauss and Joseph Cropsey (Chicago: University of Chicago Press, 1987), 421; see also Kennington, "Interpretive Essay," 59–77 passim.

[17] Melzer, *Philosophy Between the Lines*, 15, 140, 252–53, 270–71, 308–9; Arthur M. Melzer, "Appendix: A Chronological Compilation of Testimonial Evidence for Esotericism," s.v. "René Descartes," 55–56, accessed March 21, 2020, https://www.press.uchicago.edu/sites/melzer/melzer_appendix.pdf.

[18] Descartes, *Discourse on Method*, pt. 4, 32–33; Descartes, *Meditations on First Philosophy*, First Meditation, Second Meditation, in *The Philosophical Writings of Descartes*, trans. John Cottingham, Robert Stoothoff, and Dugald Murdoch (New York: Cambridge University Press, 1984), 2:12–23, Kindle; Descartes, *Principles of Philosophy*, pt. 1, §§ 1–11, in *Philosophical Writings*, 1:193–96.

[19] Descartes, *Meditations on First Philosophy*, Sixth Meditation, in *Philosophical Writings*, 2:54. The language enclosed in dashes was added in the later French version. 2:54n77.

[20] Descartes, *Meditations on First Philosophy*, synopsis of the Third Meditation, in *Philosophical Writings*, 2:10. The language enclosed in dashes was added in the later French version. 2:10n13.

[21] See Hanna and Maiese, *Embodied Minds in Action*, 50–52, 56–57 (observing, in reference to mind-body dualism, that Descartes himself made statements suggesting that "paradoxically enough the real Descartes is *not* a card-carrying Cartesian in every one of his guises" [52, italics in the original]).

[22] See, among many other examples, Damasio, *Descartes' Error*.
[23] Leo Strauss, *Persecution and the Art of Writing*, 182–83.
[24] Strauss, 150.
[25] Descartes, *Philosophical Writings*, 1:205.
[26] Descartes, 1:205.
[27] Descartes. 1:205–6.
[28] Descartes, 1:206.
[29] Descartes, 1:206.
[30] Descartes, 1:379–81.
[31] See Descartes, *Meditations on First Philosophy*, Fourth Meditation, in *Philosophical Writings*, 2:39–43.
[32] See the section "Theological Arguments against Free Will" near the beginning of Chapter 1.
[33] Hobbes's Twelfth Objection, in Descartes, *Philosophical Writings*, 2:133–34.
[34] Descartes, *Philosophical Writings*, 2:134.
[35] Descartes, *Principles of Philosophy*, pt. 1, § 28, in *Philosophical Writings*, 1:202–3; Descartes, *The Search for Truth by Means of the Natural Light*, in *Philosophical Writings*, 2:400–20 passim.
[36] *Critique of the Power of Judgment* (1790), 5:174–76; *Critique of Pure Reason* (1787), Bxvi–xxxvii (Preface to Second Edition), B305–12 (some of the corresponding material in A was superseded by B). See the Bibliography in the present book for the Cambridge University Press editions and translations of the Kant writings referenced and quoted in the present chapter. All page citations to Kant's writing are to the customary German Academy pagination as indicated in these editions. Following standard practice, citations to the *Critique of Pure Reason* are, where appropriate, to both the first (A) and second (B) editions of that work. The original publication dates of Kant's cited writings are stated in

parentheses. Except as otherwise noted, Kant was the author of all writings cited in this section.

[37] For some of Kant's discussions of free will versus predeterminism, see *Critique of Pure Reason*, Bxxvi–xxx, A488–49/B476–77, A532–58/B560–86; *Groundwork of the Metaphysics of Morals* (1785), 4:433, 446–48, 455–63; *Critique of Practical Reason* (1788), 5:28–30, 94–106; and *Critique of Judgment*, 4:435.

[38] *Groundwork*, 4:421 (emphasis in the original); see also *Critique of Practical Reason*, 5:30, and *Metaphysics of Morals* (1797), 6:225, 376, 389.

[39] *Groundwork*, 4:425 (emphasis in the original).

[40] "Idea for a Universal History with a Cosmopolitan Aim" (1784), 8:23n (other planets); *Critique of Practical Reason*, 5:32 ("all finite beings that have reason and will and even includes the infinite being as the supreme intelligence"); *Anthropology from a Pragmatic Point of View* (1998), 7:331–33 (other planets, angels); *Metaphysics of Morals*, 6:405 ("holy (supernatural) being"), 6:435 ("seraph").

[41] *Lectures on the Philosophy of Religion* (ca. 1783–86), 28:1068; *Critique of Practical Reason*, 5:28–30, 31, 109, 133; *Religion within the Boundaries of Mere Reason* (1793), 6:49n; *Metaphysics of Morals*, 6:225.

[42] *Groundwork*, 4:463 (emphasis in the original); cf. *Critique of Practical Reason*, 5:30–32.

[43] *Metaphysics of Morals*, 6:225 (emphasis in the original).

[44] *Critique of Pure Reason*, A538–58/B566–86; *Critique of Practical Reason*, 5:94–102.

[45] Bob Doyle, "Immanuel Kant," The Information Philosopher, accessed September 19, 2020, https://informationphilosopher.com/solutions/philosophers/kant/ (italics in the original).

⁴⁶ Hanna, *Deep Freedom and Real Persons*, 89–102; Henry E. Allison, *Kant's Theory of Freedom* (Cambridge: Cambridge University Press, 1990), 47–53.

⁴⁷ Hanna, *Deep Freedom and Real Persons*, 87–102.

⁴⁸ *Lectures*, 28:1045, 1116; *Critique of Practical Reason*, 5:100–3; *Religion*, 6:52n; *Conflict of the Faculties* (1798), 7:41.

⁴⁹ *Metaphysics of Morals*, 6:289 (referring to a passage in chapter 4 of book 1 of Smith's *Wealth of Nations*); *Anthropology*, 7:209 (referring to a passage in chapter 3 of book 2 of the *Wealth of Nations*). An editorial note in the Cambridge University Press volume of Kant's writings on *Anthropology, History, and Education* (p. 489, n. 22, Kindle) cites Wolfgang Kersting, "who calls Kant's interpretation of nature 'the sister of Smith's invisible hand and the forerunner of the Hegelian cunning of reason' (*Wohlgeordnete Freiheit*, p. 85)." Kant's philosophy of history is a subtle and nuanced precursor of Hegel's and Marx's philosophies of history.

⁵⁰ *Lectures*, 28:1103–17; "Idea for a Universal History with a Cosmopolitan Aim" (1784); "On the Common Saying: That May Be Correct in Theory, but It Is of No Use in Practice" (1793), 8:307–13; *Critique of Judgment*, 5:432–34; *Religion*, 6:27; *Toward Perpetual Peace* (1795), First Supplement ("On the Guarantee of Perpetual Peace"), 8:360–68; *Conflict of the Faculties*, pt. 2 ("An Old Question Raised Again: Is the Human Race Constantly Progressing?"), 4:79–94.

⁵¹ *Metaphysics of Morals*, 6:354–55.

⁵² Standford Rives, *Did Calvin Murder Servetus?* (Charleston, SC: BookSurge, 1968), chap. 15, n. 438, loc. 10200–15 of 11565, and chap. 33, n. 838, loc. 11159–64, Kindle.

⁵³ Gregory B. Graybill, *Evangelical Free Will: Philipp Melanchthon's Doctrinal Journey on the Origins of Faith* (Oxford: Oxford University Press, 2010), 224–26.

⁵⁴ Alister E. McGrath, *Reformation Thought: An Introduction*, 4th ed. (Malden, MA: Wiley-Blackwell, 2012), 195–97, Kindle.

⁵⁵ McGrath, 203–4; see also Rives, *Calvin*, chap. 15, loc. 3619–22, 3628–29.

⁵⁶ L. DeAne Lagerquist, *The Lutherans* (Westport, CT: Greenwood, 1999), chap. 5, loc. 1607–16 of 3584, Kindle; "Brief Statement of the Doctrinal Position of the Missouri Synod," paras. 35–40, adopted 1932, accessed June 18, 2019, https://www.lcms.org/about/beliefs/doctrine/brief-statement-of-lcms-doctrinal-position#election-of-grace.

⁵⁷ Cf. Waller, "Stubborn Illusion of Moral Responsibility," 66.

⁵⁸ Doyle uses the authorial name "Bob Doyle" for his books and online website. He uses "Robert O. Doyle" in his formal academic papers.

⁵⁹ See Bob (Robert O.) Doyle, *Free Will: The Scandal in Philosophy* (Cambridge, MA: I-Phi Press, 2011), Kindle; Robert O. (Bob) Doyle, "The Two-Stage Model to the Problem of Free Will: How Behavioral Freedom in Lower Animals Has Evolved to Become Free Will in Humans and Higher Animals," in *Is Science Compatible with Free Will?*, ed. Suarez and Adams, chap. 16, Kindle, available at https://www.academia.edu/38892294/The_Two_Stage_Model_to_the_Problem_of_Free_Will_How_Behavioral_Freedom_in_Lower_Animals_Has_Evolved_to_Become_Free_Will_in_Humans_and_Higher_Animals; Bob Doyle, "Jamesian Free Will, The Two-Stage Model of William James," *William James Studies* 5 (2010): 1–28, https://www.academia.edu/38892372/Jamesian_Free_Will_the_Two_stage_Model_of_William_James; Bob Doyle, "The Cogito Model," The Information Philosopher, accessed June 30, 2021, https://informationphilosopher.com/freedom/cogito/; and his

many other essays on free will on his Information Philosopher website (http://informationphilosopher.com/freedom/).

[60] Doyle, *Free Will*, 252, 261, 325, 368, 369, 381–82; Doyle, "Two-Stage Model," 235–36; 240–41, 244, 246, 252; Doyle "Jamesian Free Will," 21–23.

[61] See generally Doyle's book *Free Will* and his article "Two-Stage Model." See also Doyle, "Jamesian Free Will," 6–9, and Doyle, "The Cogito Model."

[62] Doyle, *Free Will*, 381.

[63] Doyle, 417.

[64] Doyle, "Two-Stage Model," 242–49; Doyle, "Jamesian Free Will,"13.

[65] Doyle, "Jamesian Free Will," 12-13.

[66] William R. (W. R.) Klemm, *Making a Scientific Case for Conscious Agency and Free Will* (n.p.: Elsevier, Academic Press, 2016), Kindle; W. R. Klemm, *Atoms of Mind: The "Ghost in the Machine" Materializes* (n.p.: Springer Science+Business Media, 2011), Kindle; W. R. Klemm, *Mental Biology: The New Science of How the Brain and Mind Relate* (Amherst, NY: Prometheus, 2014), Kindle. *Atoms of Mind* is Klemm's most scholarly treatment of these matters. For additional information about his approach, see my review of *Atoms of Mind* at https://www.goodreads.com/review/show/2690140181?book_show_action=false&from_review_page=1. See also Klemm's paper on the experiments of Benjamin Libet et al. ("Free Will Debates: Simple Experiments Are Not So Simple") cited in the preceding chapter.

[67] Klemm, *Scientific Case for Free Will*, 3.

[68] Klemm, 26.

[69] Klemm, 26–27.

[70] Klemm, 98.

[71] Peter Ulric Tse, *The Neural Basis of Free Will: Criterial Causation* (Cambridge, MA: MIT Press, 2013); Tse, "Two Types of Libertarian Free Will," in Caruso and Flanagan, *Neuroexistentialism*, 162–90, Kindle. Although not discussed in the present book, he also created a set of video lectures on free will titled "Libertarian Free Will: Neuroscientific and Philosophical Evidence": https://www.youtube.com/playlist?list=PLCh78lhDREMyIOCl3-9BeOWk3Q9MtxWGv.

[72] Christian List, *Why Free Will is Real* (Cambridge, MA: Harvard University Press, 2019).

[73] Terrence W. Deacon, *Incomplete Nature: How Mind Emerged from Matter* (New York: W. W. Norton, 2013), Kindle.

[74] Henry P. Stapp, *Quantum Theory and Free Will: How Mental Intentions Translate into Bodily Actions* (Cham, SZ: Springer, 2017).

[75] Henry P. Stapp, *Mindful Universe: Quantum Mechanics and the Participating Observer*, 2nd ed. (Heidelberg: Springer, 2011); Henry P. Stapp, *Mind, Matter and Quantum Mechanics*, 3rd ed. (Berlin: Springer, 2009); Henry P. Stapp, "Quantum Reality and Mind," chap. 49 in *Consciousness and the Universe: Quantum Physics, Evolution, Brain and Mind*, ed. Roger Penrose, Stuart Hameroff, and Subhash Kak (Cambridge, MA: Cosmology Science, 2017); Henry P. Stapp, "Quantum Theory of Mind," chap. 6 in *Contemporary Dualism: A Defense*, ed. Andrea Lavazza and Howard Robinson (New York: Routledge, 2014); Henry P. Stapp, "Quantum Reality and Mind," chap. 2 in *Quantum Physics of Consciousness*, ed. Subhash Kak, Roger Penrose, and Stuart Hameroff (Cambridge, MA: Cosmology Science, 2012); Henry P. Stapp, "Attention, Inattention and Will in Quantum Physics," in *The Volitional Brain: Toward a Neuroscience of*

Free Will, ed. Benjamin Libet, Anthony Freeman, and Keith Sutherland (Exeter, UK: Imprint Academic, 1999), 143–64.

[76] Jeffrey M. Schwartz and Sharon Begley, *The Mind and the Brain: Neuroplasticity and the Power of Mental Force* (New York: HarperCollins, 2002). Sharon Begley (1956–2021) was a science journalist. *The Mind and the Brain* was based primarily on Schwartz's clinical, experimental, and research findings.

[77] Schwartz and Begley, 15.

[78] Jeffrey M. Schwartz, Henry P. Stapp, and Mario Beauregard, "Quantum Physics in Neuroscience and Psychology: A Neurophysical Model of Mind-Brain Interaction," *Philosophical Transactions: Biological Sciences* 360, no. 1458 (June 29, 2005): 1309–27, https://doi.org/10.1098/rstb.2004.1598, https://www.jstor.org/stable/30041344. A freely accessible full text of this paper is available at https://www.ncbi.nlm.nih.gov/pmc/articles/PMC1569494/.

[79] Although Schwartz evidently invented the four-step process and had other original ideas discussed in his book, he acknowledges that his colleagues also contributed to the OCD research program and protocols.

[80] The discussion near the end of chapter 10 of *The Mind and the Brain* explains the details further; see also chapter 8 ("The Quantum Brain"). Cf. Stapp, *Mindful Universe*, chap. 4.

[81] I have not read or viewed any of Schwartz's writings or videos other than the paper and book discussed in the text. Accordingly, I express no opinion regarding them.

[82] Norman Doidge, *The Brain That Changes Itself: Stories of Personal Triumph from the Frontiers of Brain Science* (New York: Penguin, 2007), 372 (note), Kindle.

[83] Norman Doidge, "Neuroplasticity, Perfectibility, and Three Ideas of Nature," in *Recovering Reason: Essays in Honor of Thomas L. Pangle*, ed. Timothy Burns (Lanham, MD:

Lexington Books, 2010), 443–62. An earlier version of a portion of this essay is set forth in Appendix 2 ("Plasticity and the Idea of Progress") of Doidge's *The Brain That Changes Itself*.

[84] Norman Doidge, *The Brain's Way of Healing: Remarkable Discoveries and Recoveries from the Frontiers of Neuroplasticity* (New York: Penguin, 2016), Kindle.

Chapter 3. Free Will and Human Nature

[1] See the subsections on Robert O. (Bob) Doyle, William R. (W. R.) Klemm, Peter Ulric Tse, Henry P. Stapp, and Jeffrey M. Schwartz in Chapter 2. In addition to the writings cited in those subsections, see Ilya Prigogine, *The End of Certainty: Time, Chaos, and the New Laws of Nature* (New York: Free Press, 1997); Stuart A. Kauffman, *The Origins of Order: Self-Organization and Selection in Evolution* (New York: Oxford University Press, 1993); Stuart A. Kauffman, *A World Beyond Physics: The Emergence and Evolution of Life* (New York: Oxford University Press, 1999), Kindle; Deacon, *Incomplete Nature*; Jeremy Sherman, *Neither Ghost nor Machine: The Emergence and Nature of Selves* (New York: Columbia University Press, 2017), Kindle; Brian G. Henning and Adam C. Scarfe, eds., *Beyond Mechanism: Putting Life Back into Biology* (Lanham, MD: Lexington Books, 2013); Hanna, *Deep Freedom and Real Persons*; Hanna and Maiese, *Embodied Minds in Action*; Roger Penrose, Stuart Hameroff, and Subhash Kak, eds., *Consciousness and the Universe: Quantum Physics, Evolution, Brain and Mind* (Cambridge, MA: Cosmology Science, 2017), Kindle; and Alexander Wendt, *Quantum Mind and Social Science: Unifying Physical and Social Ontology* (Cambridge: Cambridge University Press, 2015), Kindle. Cf. Nancy Cartwright, *How the Laws of Physics Lie* (Oxford: Clarendon, 1983); Nancy Cartwright and

Keith Ward, eds., *Rethinking Order: After the Laws of Nature* (London: Bloomsbury Academic, 2016), Kindle; Steven Horst, *Laws, Mind, and Free Will* (Cambridge, MA: MIT Press, 2011); and Sabine Hossenfelder, *Lost in Math: How Beauty Leads Physics Astray* (New York: Basic Books, 2018), Kindle. I do not agree with every position taken by these authors, nor do they agree with each other on every issue. I cite these writings only in support of the points made in the referencing paragraph in the text. Interested readers can study and independently evaluate these and other such works.

[2] Kauffman, *A World Beyond Physics*, 11.

[3] Kauffmann, 111–12.

[4] Kauffman, 11.

[5] Aristotle, *Metaphysics* 1013a24–36, in *Aristotle's "Metaphysics"*, trans. Joe Sachs, 78. Aristotle gives the same account in *Physics* 194b24–35: see *Aristotle's "Physics": A Guided Study*, trans. Joe Sachs (New Brunswick, NJ: Rutgers University Press, 1995), 54, 57–58. Sachs's literal translation of and wise commentary on these passages are excellent. The account of Aristotle's four causes in Wikipedia is also, surprisingly, helpful. Wikipedia, s.v. "Four Causes," last modified June 18, 2021 13:13, https://en.wikipedia.org/wiki/Four_causes.

[6] Francis Bacon, *The Advancement of Learning*, bk. 2, § 7, ¶ 5 (1605), in Francis Bacon, *The Advancement of Learning and New Atlantis*, ed. Thomas Case (London: Oxford University Press, 1906), 110–11; Thomas Hobbes, *De Corpore*, pt. 2, chap. 9, ¶ 4, and chap 10, ¶ 7 (1656), in *Body, Man, and Citizen: Selections from Thomas Hobbes*, ed. Richard S. Peters (New York: Collier Books, 1962), 116, 124.

[7] Cees Leijenhors, "Hobbes's Theory of Causality and Its Aristotelian Background," *The Monist* 79, no. 3 (July 1996): 426, https://www.jstor.org/stable/27903492.

[8] Kauffman, *A World Beyond Physics*, chaps. 2, 8, 11; Deacon, *Incomplete Nature*, esp. 54–61.

[9] For discussions of various permutations of compatibilism, see McKenna, "Compatibilism," §§ 1.3–4.5; John Martin Fisher, "Compatibilism" (chap. 2) and "Response to Kane, Pereboom, and Vargas" (chap. 6), in *Four Views on Free Will* (Madden, MA: Blackwell, 2007), by John Martin Fischer, Robert Kane, Derk Pereboom, and Manuel Vargas; and Kane, *Oxford Handbook of Free Will*, 2nd ed., pts. 4–5 (essays by various authors regarding compatibilism/soft determinism).

[10] See, for example, the essays collected in part 5 of Kane, *Oxford Handbook of Free Will*, 2nd ed.

[11] Clark, "Experience and Autonomy," 252 (emphasis added).

[12] Plato, *Apology of Socrates* 21d (my translation).

[13] See the subsections on Robert O. (Bob) Doyle, William R. (W. R.) Klemm, and Peter Ulric Tse in Chapter 2.

[14] Prigogine, *The End of Certainty*; Kauffman, *The Origins of Order*; Kauffmann, *A World Beyond Physics*.

[15] Kauffman, *The Origins of Order*; Kauffmann, *A World Beyond Physics*; Thomas Nagel, *Mind and Cosmos: Why the Materialist Neo-Darwinian Conception of Nature Is Almost Certainly False* (New York: Oxford University Press, 2012). See also Robert Hanna, "Nagel and Me: Beyond the Scientific Conception of the World," n.d., https://www.academia.edu/4348336/Nagel_and_Me_Beyond_the_Scientific_Conception_of_the_World. Hanna addresses, among other things, the mind-body problem and the issue of consciousness, subjects that are also treated at length in Nagel's *Mind and Cosmos*. These are huge questions, to which Hanna and his coauthor Michelle Maise devoted an entire book, *Embodied Minds in Action*. To discuss such matters in detail in the present book would require a long detour. Suffice it to say that I agree with Hanna, Maise, Nagel, and several other authors cited in this chapter (and

many I haven't cited) that subjective consciousness exists and has not been satisfactorily explained by physical processes alone. Note, in this connection, the quotation from Thomas W. Clark in the preceding section of this chapter and the discussion of consciousness in the subsection on Jeffrey M. Schwartz in Chapter 2.

BIBLIOGRAPHY

Allison, Henry E. *Kant's Theory of Freedom*. Cambridge: Cambridge University Press, 1990.

Aristotle. *Eudemian Ethics*. Edited and translated by Brad Inwood and Raphael Woolf. Cambridge: Cambridge University Press, 2013.

Aristotle. *Eudemian Ethics*. In *Aristotle: The "Athenian Constitution," The "Eudemian Ethics," and "On Virtues and Vices"*, Greek text with facing English translation by H. Rackham, 198–477. Cambridge, MA: Harvard University Press, 1952.

Aristotle. *Metaphysics*. In *Aristotle's "Metaphysics"*, translated and edited by Joe Sachs. Santa Fe, NM: Green Lion Press, 2002.

Aristotle. *Nicomachean Ethics*. In *Aristotle's "Nicomachean Ethics"*, translated with an interpretive essay, notes, and glossary by Robert C. Bartlett and Susan D. Collins. Chicago: University of Chicago Press, 2011.

Aristotle. *Physics*. In *Aristotle's "Physics": A Guided Study*, translated with introduction, commentary, and explanatory glossary by Joe Sachs. New Brunswick, NJ: Rutgers University Press, 1995.

Arp, Robert, Steven Barbone, and Michael Bruce, eds. *Bad Arguments: 100 of the Most Important Fallacies in Western Philosophy*. Hoboken, NJ: Wiley-Walton, 2018. Kindle.

Augustine of Hippo. *A Treatise on the Predestination of the Saints*. Translated by Philip Schaff. In Augustine, *The Collected Works of 26 Essential Treatises*, edited by Philip Schaff, loc. 17403–18530 of 44835, Kindle. Amazon Digital Services, 2011.

Augustine of Hippo. *On Free Choice of the Will*. Translated by Thomas Williams. Indianapolis: Hackett, 1993. Kindle.

Augustine of Hippo. *Reconsiderations*. In Augustine, *On Free Choice of the Will*, 124–29.

Bacon, Francis. *The Advancement of Learning and New Atlantis*. Edited by Thomas Case. London: Oxford University Press, 1906.

Balaguer, Mark. *Free Will*. Cambridge, MA: MIT Press, 2014.

Bayne, Tim. "Libet and the Case for Free Will Scepticism." In *Free Will and Modern Science*, edited by Richard Swinburne, chapter 2. New York: Oxford University Press, 2011.

Bayne, Tim. "Phenomenology and the Feeling of Doing: Wegner on the Conscious Will." In Pockett, *Does Consciousness Cause Behavior?*, chapter 9.

Blackmore, Susan. "Living without Free Will." In Caruso, *Exploring the Illusion of Free Will*, chapter 9.

Brigandt Ingo, and Alan Love. "Reductionism in Biology." The Stanford Encyclopedia of Philosophy. Spring 2017 ed. Edited by Edward N. Zalta. https://plato.stanford.edu/archives/spr2017/entries/reduction-biology/.

Burns, Timothy, ed. *Recovering Reason: Essays in Honor of Thomas L. Pangle*. Lanham, MD: Lexington Books, 2010.

Calvin, John. "Articles concerning Predestination." In *Calvin: Theological Treatises*, 178–80. Philadelphia: Westminster, 1954.

Calvin, John. *Institutes of the Christian Religion*. Translated by Henry Beveridge. 1845. Reprint, Grand Rapids, MI: Wm. B. Eerdmans, 2001.

Calvin, John. *Theological Treatises*. Translated and edited by J. K. S. Reid. Philadelphia: Westminster, 1954.

Cartwright, Nancy. *How the Laws of Physics Lie*. Oxford: Clarendon, 1983.

Cartwright, Nancy, and Keith Ward, eds. *Rethinking Order: After the Laws of Nature*. London: Bloomsbury Academic, 2016. Kindle.

Caruso, Gregg D, ed. *Exploring the Illusion of Free Will and Moral Responsibility*. Lanham, MD: Lexington Books, 2013. Kindle.

Caruso, Gregg D. *Free Will and Consciousness: A Determinist Account of the Illusion of Free Will*. Lanham, MD: Lexington Books, 2012.

Caruso, Gregg D. "Introduction." In Caruso, *Exploring the Illusion of Free Will*.

Clark, Thomas W. "Experience and Autonomy: Why Consciousness Does and Does Not Matter." In Caruso, *Exploring the Illusion of Free Will*, chapter 13.

Couenhoven, Jesse. *Predestination: A Guide for the Perplexed*. London: T&T Clark, 2018. Kindle.

Culver, Jennifer. "False Dilemma." In Arp Barbone, and Bruce, *Bad Arguments*, chapter 81.

Damasio, Antonio. *Descartes' Error: Emotion, Reason, and the Human Brain*. New York: Penguin Books, 1994. Kindle.

Deacon, Terrence W. *Incomplete Nature: How Mind Emerged from Matter*. New York: W. W. Norton, 2013. Kindle.

Dennett, Daniel C. *Brainstorms: Philosophical Essays on Mind and Psychology*. 40th anniversary ed. Cambridge, MA: MIT Press, 2017. Kindle.

Dennett, Daniel C. *Elbow Room: The Varieties of Free Will Worth Having*. New edition. Cambridge, MA: MIT Press, 2015.

Dennett, Daniel C. *Freedom Evolves*. New York: Penguin, 2004. Kindle.

Descartes, René. *Discourse on Method*. Translated by Richard Kennington, edited by Pamela Kraus and Frank Hunt. Newburyport, MA: Focus, 2007.

Descartes, René. *Meditations on First Philosophy*. In *The Philosophical Writings of Descartes*, 2:1–62.

Descartes, René. *Principles of Philosophy*. In *The Philosophical Writings of Descartes*, 1:177–292.

Descartes, René. *The Philosophical Writings of Descartes*. Vol. 1. Translated by John Cottingham, Robert Stoothoff, and

Dugald Murdoch. New York: Cambridge University Press, 1985. Kindle.

Descartes, René. *The Philosophical Writings of Descartes*. Vol. 2. Translated by John Cottingham, Robert Stoothoff, and Dugald Murdoch. New York: Cambridge University Press, 1984.

Descartes, René. *The Philosophical Writings of Descartes*. Vol. 3. Translated by John Cottingham, Robert Stoothoff, Dugald Murdoch, and Anthony Kenny. New York: Cambridge University Press, 1991.

Descartes, René. *The Search for Truth by Means of the Natural Light*. In *The Philosophical Writings of Descartes*, 2:400–20.

Doidge, Norman. "Neuroplasticity, Perfectibility, and Three Ideas of Nature." In Burns, *Recovering Reason*, 443–62.

Doidge, Norman. *The Brain That Changes Itself: Stories of Personal Triumph from the Frontiers of Brain Science*. New York: Penguin, 2007, Kindle.

Doidge, Norman. *The Brain's Way of Healing: Remarkable Discoveries and Recoveries from the Frontiers of Neuroplasticity*. New York: Penguin, 2016, Kindle.

Doyle, Bob (Robert O.). "Benjamin Libet." The Information Philosopher. https://informationphilosopher.com/solutions/scientists/libet/.

Doyle, Bob (Robert O.). "Comprehensive Compatibilism." The Information Philosopher. https://informationphilosopher.com/freedom/comprehensive_compatibilism.html.

Doyle, Bob (Robert O.). "Emergent Determinism." The Information Philosopher. http://informationphilosopher.com/freedom/emergent_determinism.html.

Doyle, Bob (Robert O.). *Free Will: The Scandal in Philosophy*. Cambridge, MA: I-Phi Press, 2011. Kindle.

Doyle, Bob (Robert O.). "Jamesian Free Will, The Two-Stage Model of William James." *William James Studies* 5 (2010): 1–28.

https://www.academia.edu/38892372/Jamesian_Free_Will_the_Two_stage_Model_of_William_James;

Doyle, Bob (Robert O.). "John-Dylan Haynes." The Information Philosopher. https://informationphilosopher.com/solutions/scientists/haynes/.

Doyle, Bob (Robert O.). "The Cogito Model." The Information Philosopher. https://informationphilosopher.com/freedom/cogito/;

Doyle, Bob (Robert O.). The Information Philosopher (section of website on free will). https://informationphilosopher.com/freedom/.

Doyle, Robert O. (Bob). "The Two-Stage Model to the Problem of Free Will: How Behavioral Freedom in Lower Animals Has Evolved to Become Free Will in Humans and Higher Animals." In Suarez, and Adams, *Is Science Compatible with Free Will?*, chapter 16.

Einstein, Alfred. *Mozart: His Character, His Work*. Translated by Arthur Mendel and Nathan Broder. New York: Oxford University Press, 1945.

Fearnside, W. Ward, and William B. Holther. *Fallacy: The Counterfeit of Argument*. Englewood Cliffs, NJ: Spectrum, 1959.

Fischer, John Martin. "Compatibilism." In Fischer, Martin, Kane, Pereboom, and Vargas, *Four Views on Free Will*, chapter 2.

Fischer, John Martin. "Response to Kane, Pereboom, and Vargas." In Fischer, Martin, Kane, Pereboom, and Vargas, *Four Views on Free Will*, chapter 6.

Fischer, John Martin, Robert Kane, Derk Pereboom, and Manuel Vargas. *Four Views on Free Will*. Madden, MA: Blackwell, 2007.

Fitzgerald, F. Scott. *The Great Gatsby*. New York: Scribner, 1925. Reprinted with preface, notes, and corrections by Matthew J. Bruccoli. New York: Scribner Paperback Fiction, 1992. Page references are to the 1992 edition.

Freeman, Walter J. *How Brains Make Up Their Minds*. New York: Columbia University Press, 2000.

Graybill, Gregory B. *Evangelical Free Will: Philipp Melanchthon's Doctrinal Journey on the Origins of Faith*. Oxford: Oxford University Press, 2010.

Hanna, Robert. *Deep Freedom and Real Persons—A Study in Metaphysics*. New York: Nova Science, 2018. Volume 2 of Robert Hanna, *The Rational Human Condition*.

Hanna, Robert. "Nagel and Me: Beyond the Scientific Conception of the World." N.d. https://www.academia.edu/4348336/Nagel_and_Me_Beyond_the_Scientific_Conception_of_the_World.

Hanna, Robert, and Michelle Maiese. *Embodied Minds in Action*. Oxford: Oxford University Press, 2009.

HarperCollins Study Bible. Rev. ed. Edited by Harold W. Attridge and Wayne A. Meeks. New York: HarperCollins, 2006. Kindle.

Harris, Sam. *Free Will*. New York: Free Press, 2012. Kindle.

Hasker, William. "Divine Knowledge and Human Freedom." In Kane, *Oxford Handbook of Free Will*, 2nd ed., chapter 2.

Hawking, Stephen. "Is Everything Determined?" In *Black Holes and Baby Universes and Other Essays*, chapter 12. New York: Bantam Books, 1994.

Haynes, John-Dylan, and Michael Pauen. "The Complex Network of Intentions." In Caruso, *Exploring the Illusion of Free Will*, chapter 12.

Henning, Brian G., and Adam C. Scarfe, eds. *Beyond Mechanism: Putting Life Back into Biology*. Lanham, MD: Lexington Books, 2013.

Hobbes, Thomas. *Body, Man, and Citizen: Selections from Thomas Hobbes*, Edited by Richard S. Peters. New York: Collier Books, 1962.

Hobbes, Thomas. *Hobbes and Bramhall on Liberty and Necessity*. Edited by Vere Chappell. Cambridge: Cambridge University Press, 1999.

Hobbes, Thomas. *Leviathan*. Revised student ed. Edited by Richard Tuck. Cambridge: Cambridge University Press, 1996.

Hobbes, Thomas. Objections to Descartes's *Meditations on First Philosophy* with Descartes's replies. In *The Philosophical Writings of Descartes*, 2:121–37.

Hobbes, Thomas. *Of Liberty and Necessity*. In *Hobbes and Bramhall*, 15–42.

Hobbes, Thomas. *The Questions concerning Liberty, Necessity, and Chance*. In *Hobbes and Bramhall*, 69–90.

Hoefer, Carl. "Causal Determinism." The Stanford Encyclopedia of Philosophy. Spring 2016 ed. Edited by Edward N. Zalta. https://plato.stanford.edu/entries/determinism-causal/.

Honderich, Ted. "Determinism, Compatibilism and Incompatibilism, Actual Consciousness and Subjective Physical Worlds, Humanity." In Caruso, *Exploring the Illusion of Free Will*, 54.

Honderich, Ted. "Effects, Determinism, Neither Compatibilism nor Incompatibilism, Consciousness." In Kane, *Oxford Handbook of Free Will*, 2nd ed., 442–48.

Honderich, Ted. *How Free Are You: The Determinism Problem*. 2nd ed. Oxford: Oxford University Press, 2002.

Horst, Steven. *Laws, Mind, and Free Will*. Cambridge, MA: MIT Press, 2011.

Hossenfelder, Sabine. *Lost in Math: How Beauty Leads Physics Astray*. New York: Basic Books, 2018. Kindle.

Hume, David. *A Treatise of Human Nature: Being an Attempt to Introduce the Experimental Method of Reasoning into Moral Subjects*. 3 vols. London, 1739-40.

Hume, David. *Hume: An Enquiry Concerning Human Understanding and Other Writings*. Edited by Stephen Buckle. Cambridge: Cambridge University Press, 2007. Kindle.

James, William. "The Dilemma of Determinism." In *The Will to Believe: And Other Essays in Popular Philosophy*. New York: Longmans, Green, 1912. Kindle.

Jeannerod, Marc. "Consciousness of Action as an Embodied Consciousness." In Pockett, *Does Consciousness Cause Behavior?*, chapter 2.

Johnson, Alan E. Review of W. J. Klemm, *Atoms of Mind: The "Ghost in the Machine" Materializes* (N.p.: Springer Science+Business Media, 2011). Goodreads.com. https://www.goodreads.com/review/show/2690140181?book_show_action=false&from_review_page=1.

Kane, Robert. *A Contemporary Introduction to Free Will*. New York: Oxford University Press, 2005.

Kane, Robert, ed. *The Oxford Handbook of Free Will*. 2nd edition. New York: Oxford University Press, 2011.

Kane, Robert. *The Significance of Free Will*. New York: Oxford University Press, 1998.

Kant, Immanuel. *Anthropology from a Pragmatic Point of View*. In *Anthropology, History, and Education*, 227–429.

Kant, Immanuel. *Anthropology, History, and Education*. Edited by Günter Zöller, Robert B. Louden, Mary Gregor, Paul Guyer, Holly Wilson, Allen W. Wood, and Arnulf Zweig. New York: Cambridge University Press, 2007. Kindle.

Kant, Immanuel. *Conflict of the Faculties*. In *Religion and Rational Theology*, loc. 5687–7548 of 13325, Kindle.

Kant, Immanuel. *Critique of Practical Reason*. In *Practical Philosophy*, 133–271.

Kant, Immanuel. *Critique of Pure Reason*. Translated and edited by Paul Guyer and Allen W. Wood. New York: Cambridge University Press, 1998.

Kant, Immanuel. *Critique of the Power of Judgment*. Edited by Paul Guyer, translated by Paul Guyer and Eric Matthews. New York: Cambridge University Press, 2000.

Kant, Immanuel. *Groundwork of the Metaphysics of Morals*. In *Practical Philosophy*, 37–108.

Kant, Immanuel. "Idea for a Universal History with a Cosmopolitan Aim." In *Anthropology, History, and Education*, 107–21.

Kant, Immanuel. *Lectures on the Philosophical Doctrine of Religion.* Translated by Allen W. Wood. In *Religion and Rational Theology,* loc. 7585–10275 of 13325, Kindle.

Kant, Immanuel. "On the Common Saying: That May Be Correct in Theory, but It Is of No Use in Practice." In *Practical Philosophy,* 273–309.

Kant, Immanuel. *Practical Philosophy.* Translated and edited by Mary J. Gregor. Cambridge: Cambridge University Press, 1996.

Kant, Immanuel. *Religion and Rational Theology.* Translated and edited by Allen W. Wood and George di Giovanni. New York: Cambridge University Press, 1996.

Kant, Immanuel. *Religion within the Boundaries of Mere Reason.* In *Religion and Rational Theology,* loc. 1262–5366 of 13325, Kindle.

Kant, Immanuel. *The Metaphysics of Morals.* In *Practical Philosophy,* 353–603.

Kant, Immanuel. *Toward Perpetual Peace.* In *Practical Philosophy,* 311–51.

Kauffman, Stuart A. *A World Beyond Physics: The Emergence and Evolution of Life.* New York: Oxford University Press, 1999. Kindle.

Kauffman, Stuart A. *The Origins of Order: Self-Organization and Selection in Evolution.* New York: Oxford University Press, 1993.

Kennington, Richard. "Interpretive Essay: Descartes's *Discourse on Method.*" In Descartes, *Discourse on Method,* 59–76.

Kennington, Richard. "René Descartes." In *History of Political Philosophy,* 3rd ed., ed. Leo Strauss and Joseph Cropsey, 421–39. Chicago: University of Chicago Press, 1987.

Klemm, W. R. (William R.). *Atoms of Mind: The "Ghost in the Machine" Materializes.* N.p.: Springer Science+Business Media, 2011. Kindle.

Klemm, W. R. (William R.). "Free Will Debates: Simple Experiments Are Not So Simple." *Advances in Cognitive Psychology* 6 (2010): 47–65. https://doi.org/10.2478/v10053-008-0076-2.

Klemm, W. R. (William R.). *Mental Biology: The New Science of How the Brain and Mind Relate.* Amherst, NY: Prometheus, 2014. Kindle.

Klemm, William R. (W. R.). *Making a Scientific Case for Conscious Agency and Free Will.* N.p.: Elsevier, Academic Press, 2016). Kindle.

LaBossiere, Michael C. *76 Fallacies.* Amazon Digital Services, 2012. Kindle.

LaFave Wayne R., and Austin W. Scott Jr. *Criminal Law.* St. Paul, MN: West, 1972.

Lagerquist, L. DeAne. *The Lutherans.* Westport, CT: Greenwood, 1999.

Laplace, Pierre Simon. *A Philosophical Essay on Probabilities.* Translated (from the 6th French edition published in 1814) by Frederick Wilson Truscott and Frederick Lincoln Emory. New York: John Wiley and Sons, 1902.

Lavazza, Andrea. "Why Cognitive Sciences Do Not Prove That Free Will Is an Epiphenomenon." *Frontiers in Psychology* 10, no. 326 (2019): 1–11. https://www.academia.edu/38462524/Why_Cognitive_Sciences_Do_Not_Prove_That_Free_Will_Is_an_Epiphenomenon?email_work_card=view-paper.

Lavazza, Andrea, and Howard Robinson, eds. *Contemporary Dualism: A Defense.* New York: Routledge, 2014. Kindle.

Leijenhors, Cees. "Hobbes's Theory of Causality and Its Aristotelian Background." *The Monist* 79, no. 3 (July 1996): 426. https://www.jstor.org/stable/27903492.

Levy, Neil. "Be a Skeptic, Not a Metaskeptic." In Caruso, *Exploring the Illusion of Free Will,* chapter 5.

Libet, Benjamin. *Mind Time: The Temporal Factor in Consciousness.* Cambridge, MA: Harvard University Press, 2004.

Libet, Benjamin, Anthony Freeman, and Keith Sutherland, eds. *The Volitional Brain: Toward a Neuroscience of Free Will.* Exeter, UK: Imprint Academic, 1999.

List, Christian. *Why Free Will is Real.* Cambridge, MA: Harvard University Press, 2019.

Luther, Martin. "Preface to the Epistle of St. Paul to the Romans." Translated by Bertram Lee Woolf. In *Martin Luther: Selections from His Writings*, ed. John Dillenberger, 19–34. New York: Anchor Books, 1962. Originally published in German in 1522.

Luther, Martin. *The Bondage of the Will*. Translated by J. I. Packer and O. R. Johnston. Grand Rapids, MI: Fleming H. Revell, 1957. Originally published in German in 1525.

McGrath, Alister E. *Reformation Thought: An Introduction*. 4th ed. Malden, MA: Wiley-Blackwell, 2012.

McKenna, Michael. "Compatibilism." The Stanford Encyclopedia of Philosophy. Spring 2021 ed. Edited by Edward N. Zalta. https://plato.stanford.edu/archives/spr2021/entries/compatibilism/.

Mele, Alfred R. *Effective Intentions: The Power of Conscious Will*. New York: Oxford University Press, 2009. Kindle.

Mele, Alfred R. *Free: Why Science Hasn't Disproved Free Will*. New York: Oxford University Press. Kindle.

Melzer, Arthur M. "Appendix: A Chronological Compilation of Testimonial Evidence for Esotericism," s.v. "René Descartes." https://www.press.uchicago.edu/sites/melzer/melzer_appendix.pdf.

Melzer, Arthur M. *Philosophy Between the Lines: The Lost History of Esoteric Writing*. Chicago: University of Chicago Press, 2014. Kindle.

Mendelson, Michael. "Saint Augustine." The Stanford Encyclopedia of Philosophy. Winter 2018 ed. Edited by Edward N. Zalta. https://plato.stanford.edu/archives/win2018/entries/augustine.

Miller, Jeff, Peter Shepherdson, and Judy Trevena. "Effects of Clock Monitoring on Electroencephalographic Activity: Is Unconscious Movement Initiation an Artifact of the Clock?" *Psychological Science* 22, no. 1 (January 2011): 103–9, https://www.jstor.org/stable/40984614.

Missouri Synod (of the Lutheran Church). "Brief Statement of the Doctrinal Position of the Missouri Synod." 1932. https://www.lcms.org/about/beliefs/doctrine/brief-statement-of-lcms-doctrinal-position#election-of-grace.

Muniz, Michael J. "Hasty Generalization." In Arp, Barbone, and Bruce, *Bad Arguments,* chapter 84

Nagel, Thomas. *Mind and Cosmos: Why the Materialist Neo-Darwinian Conception of Nature Is Almost Certainly False.* New York: Oxford University Press, 2012.

Nahmias, Eddy. "Free Will and Responsibility." Wiley Interdisciplinary Reviews: Cognitive Science (2012). https://www.academia.edu/41577478/Free_will_and_responsibility.

Paul the Apostle. Letter to the Ephesians. In *HarperCollins Study Bible*, loc. 121422–948 of 129742, Kindle.

Paul the Apostle. Letter to the Romans. In *HarperCollins Study Bible*, loc. 117101–8417 of 129742, Kindle.

Pelagius. *The Letters of Pelagius: Celtic Soul Friend.* Edited by Robert Van de Weyer. Worcestershire, UK: Arthur James Ltd., 1995.

Penrose, Roger, Stuart Hameroff, and Subhash Kak, eds. *Consciousness and the Universe: Quantum Physics, Evolution, Brain and Mind.* Cambridge, MA: Cosmology Science, 2017.

Pereboom, Derk. *Free Will, Agency, and Meaning in Life.* Oxford: Oxford University Press, 2014. Kindle.

Plato. *Apology of Socrates.* In *"Plato: "Euthyphro," "Apology," "Crito," "Phaedo," "Phaedrus"*, Greek text with facing English translation by Harold North Fowler, 68–145. Cambridge, MA: Harvard University Press, 1914.

Plato. *Apology of Socrates.* In *Four Texts on Socrates*, rev. ed., translated with notes by Thomas G. West and Grace Starry West, 63–97. Ithaca, NY: Cornell University Press, 1998.

Plato. *The Republic.* In *"The Republic" of Plato*, 2nd ed., translated with notes, interpretive essay, and a new introduction by Allan Bloom. New York: Basic Books, 1991.

Pockett, Susan. "The Neuroscience of Movement." In Pockett, *Does Consciousness Cause Behavior?*, chapter 1.

Pockett, Susan, William P. Banks, and Shaun Gallagher, eds. *Does Consciousness Cause Behavior?* Cambridge, MA: MIT Press, 2006.

Prigogine, Ilya. *The End of Certainty: Time, Chaos, and the New Laws of Nature.* New York: Free Press, 1997.

Rives, Standford. *Did Calvin Murder Servetus?* Charleston, SC: BookSurge, 1968.

Schirber, Michael. "The Chemistry of Life: The Human Body." Live Science. April 16, 2009. https://www.livescience.com/3505-chemistry-life-human-body.html.

Schuessler, Jennifer. "Philosophy That Stirs the Waters." *New York Times*. April 29, 2013. https://www.nytimes.com/2013/04/30/books/daniel-dennett-author-of-intuition-pumps-and-other-tools-for-thinking.html?searchResultPosition=1.

Schwartz, Jeffrey M., and Sharon Begley *The Mind and the Brain: Neuroplasticity and the Power of Mental Force.* New York: HarperCollins, 2002.

Schwartz, Jeffrey M., Henry P. Stapp, and Mario Beauregard. "Quantum Physics in Neuroscience and Psychology: A Neurophysical Model of Mind-Brain Interaction." *Philosophical Transactions: Biological Sciences* 360, no. 1458 (June 29, 2005): 1309–27. https://www.jstor.org/stable/30041344.

Shakespeare. *Julius Caesar.* In *The First Folio of Shakespeare*, 717–38.

Shakespeare. *Macbeth.* In *The First Folio of Shakespeare*, 739–59.

Shakespeare. *The First Folio of Shakespeare: The Norton Facsimile.* 2nd ed. Edited by Charles Hinman and Peter W. M. Blayney. New York: W.W. Norton & Co., 1996.

Sherman, Jeremy. *Neither Ghost nor Machine: The Emergence and Nature of Selves.* New York: Columbia University Press, 2017. Kindle.

Smilanksy, Saul. "Free Will as a Case of 'Crazy' Ethics." In Caruso, *Exploring the Illusion of Free Will*, chapter 4.

Smith, Adam. *An Inquiry into the Nature and Causes of the Wealth of Nations*. Edited by Edwin Cannan, with an introduction by Max Lerner. New York: Modern Library, 1937.

Sorabji, Richard. *Necessity Cause and Blame: Perspectives on Aristotle's Theory*. London: Duckworth, 1983.

Stapp, Henry P. "Attention, Inattention and Will in Quantum Physics." In Libet, Freeman, and Sutherland, *The Volitional Brain*, 143–64.

Stapp, Henry P. *Mind, Matter and Quantum Mechanics*. 3rd ed. Berlin: Springer, 2009.

Stapp, Henry P. *Mindful Universe: Quantum Mechanics and the Participating Observer*. 2nd ed. Heidelberg: Springer, 2011.

Stapp, Henry P. "Quantum Reality and Mind." In Penrose, Hameroff, and Kak, *Consciousness and the Universe*, chapter 49.

Stapp, Henry P. *Quantum Theory and Free Will: How Mental Intentions Translate into Bodily Actions*. Cham, SZ: Springer, 2017.

Stapp Henry P. "Quantum Theory of Mind." In Lavazza and Robinson, *Contemporary Dualism*, chapter 6.

Steward, Helen. *A Metaphysics for Freedom*. Oxford: Oxford University Press, 2012.

Strauss, Leo. *Persecution and the Art of Writing*. Chicago: University of Chicago Press, 1952.

Strauss, Leo. *The Political Philosophy of Hobbes: Its Basis and Genesis*. Translated by Elsa M. Sinclair. Chicago: University of Chicago Press, 1952.

Strawson, Galen. "The Impossibility of Ultimate Responsibility?" In Caruso, *Exploring the Illusion of Free Will*, chapter 2.

Suarez, Antoine, and Peter Adams, eds. *Is Science Compatible with Free Will?: Exploring Free Will and Consciousness in the Light of Quantum Physics and Neuroscience*. New York: Springer Science+Business Media, 2013.

Tse, Peter Ulric. "Libertarian Free Will: Neuroscientific and Philosophical Evidence" (video lectures). https://www.youtube.com/playlist?list=PLCh78lhDREMyIOCl3-9BeOWk3Q9MtxWGv.

Tse, Peter Ulric. *The Neural Basis of Free Will: Criterial Causation*. Cambridge, MA: MIT Press, 2013.

Tse, Peter Ulric. "Two Types of Libertarian Free Will Are Realized in the Human Brain." In *Neuroexistentialism: Meaning, Morals, and Purpose in the Age of Neuroscience*, ed. Gregg D. Caruso and Owen Flanagan, chapter 16. New York: Oxford University Press, 2018.

Van de Weyer, Robert. "Introduction." In *The Letters of Pelagius*, unpaged.

Vargas, Manuel. "If Free Will Does Not Exist, Neither Does Water." In Caruso, *Exploring the Illusion of Free Will*, chapter 10.

Waller, Bruce. "The Stubborn Illusion of Moral Responsibility." In Caruso, *Exploring the Illusion of Free Will*, chapter 4.

Walton, Douglas. *Informal Logic: A Pragmatic Approach*. 2nd ed. New York: Cambridge University Press, 2008. Kindle.

Wegner, Daniel M. *The Illusion of Conscious Will*. New Edition. Boston: MIT Press, 2018. Kindle.

Wendt, Alexander. *Quantum Mind and Social Science: Unifying Physical and Social Ontology*. Cambridge: Cambridge University Press, 2015. Kindle.

Wikipedia, s.v. "Four Causes." Last modified June 18, 2021 13:13. https://en.wikipedia.org/wiki/Four_causes.

INDEX

Allison, Henry E., 63

Aristotle, 41, 48–54, 81, 87, 92, 98

Augustine of Hippo (Saint Augustine), 8–11, 15, 16, 17, 59, 63

Bacon, Francis, 92

Bartlett, Robert C., 53, 54, 55

Bayne, Tim, 112n54

Blackmore, Susan, 35–37

Bramhall, John, 13

Buffon, Comte de (Georges Louis Leclerc), 84

Calvin, John, 11–12, 14, 15, 16, 17, 59, 60, 63, 65

causal determinism. *See* predeterminism: scientific

chance. *See* indeterminism

chaos theory, 61, 90, 97, 103

Chappell, Vere, 13–14

Clark, Thomas W., 41, 95, 132–33n17

classical physics. *See* predeterminism: scientific

Collins, Susan B., 53, 54, 55

compatibilism/soft determinism: 3, 7, 15, 42–47, 50, 62, 90, 93–96, 102

complexity theory, 61, 90, 97, 103

consciousness (including the subconscious, the unconscious, and the mind-body problem), 3–5, 24–35, 39, 41, 42, 46, 48, 51, 67, 68, 72, 74–75, 98–99, 102, 132-33n17

criminal law: theories of punishment, 23–24, 36–37

Damasio, Antonio, 40

Deacon, Terrence W., 74

Dennett, Daniel C., 35, 43–47, 53, 118n111

Descartes, René, 48, 55–60, 62, 84, 87

determinism: adequate determinism, 66; ad hoc (genetic or environmental) determinism, 37–41, 90, 99–100, 102; common but incorrect term for predeterminism, 5, 20–21, 42, 44, 69, 70. *Compare* predeterminism

Doidge, Norman, 23, 84–86, 98

Doyle, Robert O. (Bob), 2–4, 17, 23, 29, 45-46, 62, 65–66, 96, 97

dualism, 48, 55–65, 84, 95, 122n21

Einstein, Albert, 1, 33, 97

fallacies, 2, 28, 32, 34, 35, 90

Fitzgerald, F. Scott: on reserving judgments, 37–39

free will: arguments against, 7–47, 90–96, 102; arguments for, 2, 48–89, 96–101, 102–3; debates about, 1–3; defined, 3–6; evolution/natural selection and, 2, 27, 42, 44, 65, 67, 68, 68, 69, 74, 91, 95, 96, 97, 102; in human life, 98–101, 103; political free will, 45, 63, 100–1

French revolution, 85

Frost, Robert: "The Road Not Taken," 88–89

Galileo, 55, 91

Hanna, Robert, 63, 118n105, 122n21, 132–33n17

Harris, Sam, 34–35

Hawking, Stephen, 7

Hobbes, Thomas, 12–17, 42–43, 59– 60, 63, 92–93, 111n43

Honderich, Ted, 20–24, 36–37

Hume, David, 43, 11718n111

indeterminism, 1–2, 5, 21–22, 42, 46–47, 65–66, 70–71, 79, 90–91, 96

Kane, Robert, 54, 118n103

Kant, Immanuel, 16–17, 43, 48, 60–63, 87

Kauffman, Stuart A., 91–92, 97

Klemm, William R. (W. R.), 4, 22–23, 39, 61, 67–68, 74, 96, 97, 127n67

Laplace, Pierre-Simon, 18–19, 39, 75

Leijenhorst, Cees, 92–93

Libet, Benjamin, 24–30, 31, 34 –35, 68, 98, 112–13n54

List, Christian, 19, 40, 73–74

Locke, John, 23–24

Luther, Martin, 11, 14, 15, 16, 17, 59, 63–65

McGrath, Alister E., 64

Melanchthon, Philipp, 64

mind–body problem. *See* consciousness

Mozart, Wolfgang Amadeus, 33

Neoplatonism, 8

neuroplasticity, 32, 77–82, 84–86, 98

Newcastle, Marquess of, 13

Newton, Isaac, 19, 75, 91–92

Newtonian physics: *See* predeterminism: scientific

Paul the Apostle (Saint Paul, Paul of Tarsus), 8, 11, 14, 15, 17, 63

Pelagius, 8–9

Plato, 50, 62, 63, 65

predeterminism: theological, 1, 7–17, 58–60, 63–65, 89–90, 102; scientific (universal natural determinism), 1–2, 5–6, 17–47, 63, 77, 90–96, 102. *Compare* determinism

Prigogine, Ilya, 61, 97

Pythagoras, 91

quantum physics/mechanics, 1–2, 19, 21–22, 27, 29, 39, 61, 66, 70, 74–79, 83–84, 90, 97, 103

relativity, 61, 97

retribution, 23–24, 36–37
revelation, 16, 60, 111–12n44
Rousseau, Jean-Jacques, 84–85
Russian revolution, 85
Schwartz, Jeffrey M., 23, 77–83, 84, 96–98
Shakespeare: *Julius Caesar*, 48; *Macbeth*, 1
social contract, 23–24
Socrates, 2, 96
soft determinism. *See* compatibilism/soft determinism
sophistry, 2–3, 17–18
Stapp, Henry P., 22, 27, 29, 61, 74–78, 83, 84, 96–97, 129n81
teleology, 92, 97, 102–3
Tse, Peter Ulric, 23, 27, 61, 68–73, 96, 97, 105–6n8, 108n18
universal natural determinism. *See* predeterminism: scientific
Wegner, Daniel M., 30–36, 98

About the Author

Alan E. Johnson, an independent philosopher and historian, is the author of *The First American Founder: Roger Williams and Freedom of Conscience* (2015), *The Electoral College: Failures of Original Intent and Proposed Constitutional and Statutory Changes for Direct Popular Vote*, Second Edition (2021), and other publications in the fields of philosophy, history, law, and political science. The present book, *Free Will and Human Life*, is the first volume of a planned philosophical trilogy on free will, ethics, and political philosophy.

Mr. Johnson holds an A.B. (political science, 1968) and an A.M. (humanities, 1971) from the University of Chicago and a J.D. (1979) from Cleveland State University, Cleveland-Marshall College of Law. He is retired from a long career as an attorney in which he focused primarily, though not exclusively, on constitutional and public law litigation.

www.ingramcontent.com/pod-product-compliance
Lightning Source LLC
Chambersburg PA
CBHW061654040426
42446CB00010B/1737